An Ape Ethic and the Question of Personhood

An Ape Ethic and the Question of Personhood

Gregory F. Tague

LEXINGTON BOOKS
Lanham • Boulder • New York • London

Published by Lexington Books
An imprint of The Rowman & Littlefield Publishing Group, Inc.
4501 Forbes Boulevard, Suite 200, Lanham, Maryland 20706
www.rowman.com

6 Tinworth Street, London SE11 5AL, United Kingdom

British Library Cataloguing in Publication Information Available

Library of Congress Cataloging-in-Publication Data

Library of Congress Control Number: 2020933479

ISBN: 978-1-7936-1970-9 (cloth)
ISBN: 978-1-7936-1971-6 (electronic)

Dedicated To
Gary L. Shapiro, who has worked over forty years on behalf of orangutans and other species to preserve their rainforest habitats

For
Fredericka Jacks and Karolina Jacks-Tague

"Write the vision, and make it plain on tablets, that he who runs may read it."

Habakkuk 2:2

Contents

Acknowledgments

I would like to thank:
The office of the Academic Dean, Jennifer Lancaster, St. Francis College, for a course release
The librarians at St. Francis College
Multiple anonymous peer reviewers

Many scholars and scientists have encouraged my pursuits in evolutionary studies, including Daniel Meyer-Dinkgräfe; David Lahti; Christopher Jensen; Carlo Alvaro; Dustin Hellberg; Shawn Thompson; Christine Webb; and Peter Canning. Supportive current and former St. Francis College colleagues include Kathleen Nolan; Kristy Biolsi; Irina Ellison; Alison Dell; Clayton Shoppa; SungHun Kim; Ian Maloney; Wendy Galgan; Gerry Galgan; Jennifer Wingate; Athena Devlin; and Timothy Houlihan. This book had humble origins in a paper for a working group on nonhuman forms of The Modern Language Association, which met over several days in January 2018, headed by Samantha Pergadia. My gratitude to Alexandra Resnick for copyediting and to Vukasin Petrovic for proofreading. Special thanks to Paul Hebner for editorial consultation.

Introduction

Breaking Boundaries

The ambition of this book is to convince readers that great apes in wooded habitats are ecosystem engineers whose practical actions are vital to the health of the world's ecosphere. Because apes have evolved cognitive mechanisms, intelligent feeding maneuvers, and social behaviors as moral individuals displaying an ethical attitude toward the land, they have earned the right to forest sovereignty. As you might imagine, this is a heavy load to lift. In making a legal argument, a worthy book like *Chimpanzee Rights* (Andrews et al., 2019) focuses mostly on the similarities between apes and humans. I, too, rely on demonstrating similarities since those capacities and abilities are a consequential part of my argument. Yet what's more important for me is to emphasize the differences between apes and humans, namely, their ability over millions of years to adapt in ways where they mesh seamlessly with their environment and sustainably manage their habitats for the livelihood and growth of other species. Whereas other books, like *Chimpanzee Rights*, might correctly focus on the advanced mental representations of apes in ways similar to humans, *An Ape Ethic* is more concerned with the eco-psychology and embodied cognition of apes which enable their agency to engage in effective ecosystem engineering. It's as if great apes view their biomes as kin and so contribute to forest health and longevity.

I need many words to elaborate the ideas in my opening paragraph. After an overview in this Introduction, chapter 1 will delineate fine points with some detail in a succession of topics. Then, the heart of the book lies in the discussion and analysis of empirical studies found in chapter 2 (on cognition and intelligence) and chapter 3 (on social behaviors), both of which might seem dense. All of this is followed by a provocative Conclusion. Let me begin with an anecdote and proceed methodically through my chapters, which indeed cover the many similarities between apes and humans, such as consciousness and free will, mental abilities, moral agency, and social behaviors. I'm hopeful that by the end of the book readers will see how my claims for similarities between humans and apes underlie my main argument for differences, where apes are the more efficacious eco-engineers, granting them, therefore, forest personhood and dominion over their tropical properties.

At a university panel on animal experimentation and neuroscience a philosopher, a research veterinarian, a bioethicist, and a neuroscientist presented their cases. The vet and the neuroscientist defended the practice of drilling holes in monkeys' heads to investigate neuronal connectivity. They further insisted that they care for the primates' welfare, though admitted how these beings are bred solely for invasive, experimental purposes. The philosopher, advocating animal rights, objected. The bioethicist was somewhat neutral. The philosopher asked why we set constraints regarding harmful research on vulnerable human populations but not on animals. The neuroscientist answered this question by asserting that he, in contrast to the philosopher, lives in the real world where the community of humans is more important than the life of any animal.

As Christine Webb et al. (2019) argue, if scientists could be more fully informed about moral philosophy (not necessarily through a bioethics panel), then they could better understand the moral status of the animals on whom they conduct experiments. My argument is a difficult one, but that's not going to stop me from attempting to change the minds, at least in part, of some people and groups. Any type of writing, including scholarship, involves risk. One puts out words and ideas from extensive reading and thinking, and yet any number of people will raise an eyebrow. My goal is to win readers over, but I understand that much of what I say, even how I say it, will arouse surprise and disbelief. That's a chance I'm willing to brush against. For instance, one could take umbrage at the neuroscientist's assertion that he more than a philosopher lives in the "real world." In fact, a large part of *An Ape Ethic* shows how the forest world of apes is far different from and more real than air conditioned offices of lab researchers. As for philosophers in the real world, one only needs to consider some of the classical thinkers, from Socrates and Epictetus, to more contemporary mind workers like Lori Gruen and Martha Nussbaum who tackle many real world issues and problems.

As you might have guessed already, this book is not without controversy. There are a number of people who will get riled up with the assertion of great ape rights and vexed by a declaration of ape personhood. My aim is not to distance readers from me but to help them agree that the territory between "animal" and human is narrow. My purpose is to unite, not to divide. Animal rights activists will criticize any argument that spotlights a prestige species. I don't include all "animals" since I want to put a face on the individual: look into the eyes of a great ape. Many scientists will hurl accusations of gross anthropomorphism. Worse, some researchers will shout that any ascription of heightened conscious awareness to "animals" is unscientific. Nonetheless, I am going to assume elevated awareness in great apes based on my extensive readings, which I will share as the book unfolds. Humanists and philosophers of dehumanization who might position humans over apes could shudder at some of my inferences (see, for example, Smith & Panaitiu, 2015). These objections, which I take very seriously, will be addressed in due course. Strong resistance to any notion of ape intelligent willfulness is reminiscent of how the Victorians, in their moral and religious strictures, were resolved not only to break but also to bring into subjection the will of a child (Houghton, 1957). My exposition is not meant simply to jar emotions but to initiate rational discussion leading to practical action that, ultimately, bears weight on the health of this planet in how we treat tropical forest bio-networks.

Many books and articles competently address aspects of great ape personhood. For instance, consult the bibliography here. Many of these arguments, though, lean heavily on ape/human resemblances. I too spend a considerable amount of time highlighting those similarities, but I do so only to infer how these two ape paths have diverged in terms of environmental care. What distinguishes my argument from others lies in the emphasis, at times subtle, on the differences between great apes and humans. Ethicist and theologian Andrew Linzey (2009) seems to suggest that animal differences in comparison with humans should not be slighted to minimize moral relevance but highlighted to sharpen our ethical feelings. Great apes are moral individuals who differ from humans, for example, in how they seem more adept as ecosystem engineers exhibiting an environmental psychology of good gestalt that surpasses what appears in most, but not all, human societies. Some people often besmirch the differences between humans and animals by incorrectly describing, Linzey would suggest, rather than glorifying how their evolved differences make them special. My analysis might be criticized as subjective. I don't think so, but if that is the case, then as one anonymous reviewer of an early form of this book has said, I don't need to make a slam-dunk argument. Readers will have to decide how convincing my argument is or not.

Another anonymous commentator of the earlier version said that any claim for ecosystem engineering is shattered by the patently obvious fact that great apes are incapable of exploiting their environments to the extent that humans have. Indeed, this is a severe objection I hope to address. This reader's objection is not so airtight since at least one chimpanzee group, as I discuss briefly later, has created some damage to the biodiversity in its parkland area. Orangutans have a voracious appetite and could conceivably deplete impoverished areas. More speculatively, part of great ape evolution, especially regarding birth intervals, might be an adaptation to secure ecological stability, which might explain why they do not need to have the know-how to destroy their kingdoms. Relatedly, some primate social structures of fission-fusion, spreading out during the day and reuniting at night, means that any given territory can accommodate a population without overcrowding, thereby limiting environmental degradation. Part of the human cycle of destruction, the agricultural and commercial industrial complex, comes back to over population. In other words, humans do not seem to exhibit some of the evolved life history controls seen in other species.

Sociologist Eileen Crist (2019) points out that Native Americans from up to 5kya did not destroy Atlantic or Pacific salmon runs and instead sustained them. On the other hand, Europeans from the Middle Ages and then European settlers in America depleted the salmon runs. What's the difference? Crist says that the indigenous people's culture was one of subsistence and reciprocity. Meanwhile some nineteenth-century thought would paint such people, or "savages," as stuck in a type of moral brutality, the current and common view of apes, or "beasts," for the uninformed. These issues will be explored and clarified in the pages that follow. It's important to consider that in spite of the cognitive and behavioral similarities of apes and humans, great apes evolved a land ethic that is far different and more ecologically friendly from anything remotely close in most "modern" human societies. That is, philosophic, scientific, and commercial modernization for humans has led to industrialization and an economy mostly based on profit and gain from and not on care for the environment.

WHY THIS BOOK

Mark Moffett (2019) makes the case that early humans like hunter-gatherers who lived in small societies dependent on wild gardens and animals were "nations." Can't we say this logic applies to great apes as well from savanna chimpanzees to Bornean orangutans? A criticism of my question might be that apes generally don't cooperate. I'll counter that in due course. For now, Moffett might point out how societies do not equate to cooperation but include various members who share a social identity. That is true of apes, including the so-called semi-solitary orangutan. This ape identity is tied to their goal of surviving in balance with their habitats as socially and ecologically conscious individuals.

Great apes demonstrate an ethic of moral individualism through many factors, which can only be outlined here and in chapter 1, but which will be treated in detail in the core chapters, 2 (cognition and intelligence) and 3 (social behaviors). For example, great apes possess and exercise advanced cognitive mechanisms that enable their intellectual and social abilities. Overall, great apes are communal creatures who consistently, whether near or far, acknowledge and at times challenge, in a type of social duel, the presence and status of others. Great apes live in garden-like ecosystems and notably have not ravaged their habitats. Can it be argued that great apes exhibit a land ethic for the well-being of their forest homes through ecosystem engineering? Rather than using broad strokes that view apes simply as another species separate from and supposedly below what is human, I consider apes as individuals and at times highlight a particular ape's unique characteristics.

There is an ape economy consisting of good land use. On physical and mental levels, consciously or not, apes seem to participate in James Lovelock's (1979) Gaia hypothesis of a self-regulating global ecosphere. Preceding modern humans, ancient hominids evolved adaptations for ecological systematics of thermostatic and population equilibrium. As Verne Grant (1963) would say, for the individual, its particular genes (phenotype) develop in response to the actions of its genotypic constitution under environmental circumstances. Genotypes can be modified by the environment into phenotypic variations which can be adaptive. Great apes are living proof of selectively inherited adaptations to survive in and maintain vast expanses of tropical forests. Furthermore, ape behaviors show that they are free and responsible individuals who tend to their ecological homes while caring for others in that environment. Not only is there an ape ethos, and more specifically an ethos for each great ape species, but there are also keenly aware individual characters evident among these species. Etymologically, the word "ethic" is related to ethos, character, and hence the title of my book.

Sue Savage-Rumbaugh (2007) interviewed "languaged" bonobos she worked with, and some of the core needs they articulated include travel, sharing information, socialization, solitude, long-term relationships, teaching young their culture, and acceptance by humans of their linguistic and mental abilities. The "languaged" bonobos Kanzi and Panbanisha, in particular, have demonstrated emergent awareness of others as well as states of advanced perception during their communication with each other and humans (Savage-Rumbaugh, 2018). Great apes are not only self-aware but are cognizant of their cultural conditions. By analogy, these sensitivities would include sharp awareness of their physical surroundings. Each great ape possesses his or her own ethical capacities of self-awareness, autonomous intentionality, freedom to act outside the control of others, and empathy. These qualities entitle great apes to be considered as persons with moral individualism.

As might be expected, I argue for great ape personhood with reference to their high level of intelligence and complex social behavior. However, my use of "personhood," a word that occurs throughout the book, is closer in meaning to person of the forest. Personhood has legal implications and those are not denied. More pointedly, the thrust of my argument is to advocate for apes, especially orangutans, as forest persons by virtue of their land ethic.

Homo sylvestris, or person of the forest, is a phrase of Nicolaes Tulp widely used after his 1641 designation and essay by the same name (van Wyhe & Kjaergaard, 2015). Citing E. O. Wilson's *Sociobiology* (1975), Richard Dawkins (1982) says science should not be advocacy but facts. My book uses empirical evidence to advocate for great ape sovereignty over their lands in line with political claims by Sue Donaldson and Will Kymlicka (2011) and Eva Meijer (2019). Specific claims for the personhood argument will include a thorough discussion of the scientific literature. However, what distinguishes my multidimensional argument is how I claim intelligence and sociality as characteristics of moral individualism. Moreover, I claim, there's an ape ethic for land evidenced through their ecological engineering.

In a nutshell, the argument follows this train of thought: Great ape forest personhood is evident by virtue of their intellectual and moral individualism because of their land ethic as ecosystem engineers. Correlated ideas, then, include the ape ethos, moral individualism, and eco-engineering. They connect since a moral individual who has a land ethic is a good steward of her shared habitat.

We know great apes are similar to us in morphology, psychology, and behavior. If I choose to act a certain way (e.g., for my own health or for the welfare of others), can we say that apes, because of genetic similarities, are also capable of making choices of "good" intent? Can we further say that because of evolved dissimilarities in a different environment of evolutionary adaptation apes experience a feeling of choice regarding a heightened sense of attention to their forest habitats? If you were to draw a three-bubble graphic of moral individualism, eco-engineering, and the land ethic, a pivot of contact for all three might be in the ape ecosystem. It might not be so much that apes consider the consequences of ecosystem engineering but experience, as neuroscientist Antonio Damasio (1999) would say, a feeling of what happens. The point is that they likely have a bodily and emotional frame of mind about what to do and so act. They are not utilitarian in this sense where concern for preferences of many count; rather, they are moral individuals whose natural activity just so happens to benefit others, whether kin, group, or unrelated beings. Ape actions contribute to the health of the forest. In this way the moral worth of an individual is not measured solely in cognitive capacity but in how the individual acts in a relationship with others in her ecosystem.

I can't prove that apes act ethically toward their land; but the evidence concerning their self-awareness, sympathetic capacities, and intelligence seems to bolster my claim that they have evolved and adapted their behavior so as not to destroy their tropical forest homes. I can't definitively prove conscious intent by apes in eco-engineering any more than one can prove the power of prayer; and yet people pray. In other words, I don't have to prove that eco-engineering is cognitively motivated. Rather, apes are motivated to tend to and survive in forests because of their evolved ethos, and why I call it a land ethic by moral individuals for ecosystem engineering. In some ways, an ape subconsciously considers, *How should I live?* and acts accordingly with some degree of care to self, others, and habitat in order to flourish. This is not to say great apes deliberately follow rational rules or procedures for consequences. Broadly speaking, they act in a meaningful way to achieve wholeness without intentional harm to others. In this way we can see how their immediate motives are conscious, but yet they don't necessarily have to demonstrate that there is a systematic, well-wrought plan to their ecosystem engineering. Instead, it's a matter of the total effect from a number of related causes, such as moral individualism for health (cognition and intelligence) and the land ethic (social behavior) for kin and community. They seem responsive to the beneficial results of their actions. This will all be explained in coming chapters.

Worth underlining is how any opinions I seem to make are actually driven by my review and discussion of the scientific data. Here's what differentiates this book from some others in three areas I attempt to interrelate:

1. *Ape ethic*: Drawing from environmentalist Aldo Leopold's (1949) call to care for the inherent and not economic value of an ecosystem, great apes exhibit a land ethic by virtue of their long, sustained evolutionary adaptive success in tropical forest habitats.
2. *Moral individualism*: Philosopher James Rachels (1990) works against sweeping utilitarian principles and instead calls for a return focus on individual identity and autonomy. Willful actions and not just consequences are important. This is not simply self-interest, though. Rather, it affirms the moral state of the individual in relation to others in a shared habitat. The assertion here is not a universal ape moral *ought* but rather an ape ethical *ability* for normative governance. While the individual has desires which can be satisfied, these are "morally" balanced against the needs of others. I apply this notion to great apes since they exhibit individual differences across cognitive and social behaviors similarly seen in human communities.
3. *Ecosystem engineering*: This refers to any animal, whether ant, worm, Arctic fox, beaver, or human. However, in my estimation great apes by virtue of their mentality, social behavior, and physical movements stimulate and sustain the living components of large-scale biotic communities with positively widespread repercussions not seen in other species. Furthermore, eco-engineering acts like a bridge between ape genes and the forests, an inclusive fitness component where environmental care by the apes is connected to their species longevity.

Readers will discover that these three ideas are intimately related in my call for ape personhood, since it is the moral individual who exhibits an ape ethic of ecosystem engineering. Readers also need to understand that lots of topics are covered, all circling around the notion of moral individualism. For example, chapter 1 provides a framework that is elaborated upon in the two analytical chapters that follow. Objections are covered as they would arise through the book. I don't want to presume that ape awareness, sentience, and sapience are self-evident. Rather, I have marshaled a substantial amount of evidence from field and lab researchers, as well as scientifically informed philosophers, to discuss and analyze an array of ape cognitive and behavioral patterns to support my claims of moral individualism tied to eco-engineering. My argument asks that sovereignty be granted to great apes as forest persons.

Bear in mind that the word "human" is an adjective. It's only proper to say human person and not simply human. Yet, why does the word "person" imply only a human? This fine point is raised since I will often say "humans" to distinguish that species from one of the great ape species. I could as easily say, for example, orangutan person. This is no small linguistic matter though somewhat outside the scope of the book. A turn to harmful anthropocentric attitudes will be visited from time to time. A great ape is an individualized *person* as much as a human being. This fact is widely accepted among primatologists, biologists, social scientists, and some philosophers. The much larger popular imagination, on the contrary, and represented in mass print or televised and social media paints apes as animals and not as persons. While the word "animal" is sometimes used here, it is not pejorative and in fact includes all moral animals, like humans and apes. Western over-emphasis on intelligence as the paramount human feature, according to Anne Russon and Kim Bard (1996), is challenged by its manifestation in characteristics of great apes. This means, say Russon and Bard, that there are no clear marks of demarcation between "human" and "animal." Animals are not commodities. Overcoming this prejudice against animals and, in my argument, apes will be the first major step in restoring the delicate balance of the earth's ecosphere.

Because many primate species are critically endangered, I hope this book will serve as a call to action for average humans who can speak and work on their behalf. People often hear scientists, social scientists, and philosophers ask a question like, *What makes us human?* Here's the answer. There is human morphological, cognitive, and social similarity to great apes from a common evolutionary history. This is so because, in the words of Matthew Mitchell and Mary Gonder (2013), *Hominidae* (the ape family) speciation occurred within a small window of time and, therefore, rendered *Homo* (the human line) "a mosaic of shared ancestry with each of the extant apes." Most researchers correctly think in terms of humans diverging from an ape-like common ancestor. However, it also stands to reason that apes might have evolved away from a human-like ancestor (Schwartz, 1987). My argument suggests as much with the acknowledgment of human/ape cognitive and behavioral relatedness but with heavy weight on the difference of apes as more capable habitat eco-engineers than humans.

ABOUT THIS BOOK

To a large extent I argue that great ape "culture" arises from the interaction of genotype with the challenges and opportunities of a tropical bio-network. This might seem obvious but is not. Ape culture consists of terrestrial stewardship from mental and social behaviors that evolved as adaptations for forest survival. Biomes are dynamic systems. Great apes and their multifaceted tropical homes have efficiently evolved together as one contiguous whole. Mutually beneficial encounters between apes regarding cultivation of trees and plants as sustainable resources suggest a land ethic culture.

This is an interdisciplinary work. Issues and subjects related to great ape personhood, ape moral individualism, ecosystem engineering, conservation efforts, and anthropogenic stress on ape habitats echo in harmony from chapter to chapter. I note for skeptics of anecdotal information how an animal expert of high standing, Marc Bekoff (2000), says that scientists need to combine the anecdotal and the empirical in an interdisciplinary manner, clearly represented in this book. While one subject under the magnifying glass is orangutans, in order to picture them fully I pull back and examine other great ape intelligence and behavior from philosophical and scientific viewpoints. This roundness of presentation means that readers will see the whole picture of great ape persecution in spite of their moral individuality as ecosystem engineers.

It's important to note how *An Ape Ethic* has a major chapter on cognitive abilities and another on social behavior. This is because I see moral individualism grounded in the areas of intelligence and sociality. Great apes are not robots; they use their intellectual autonomy for ecological engineering (a land ethic) and for social behavior (a moral ethic). These traits and tendencies, while human-like, differ in great apes when compared with most modern humans. Principally, the ape sense of community is ecologically based. The ape ethic is both inherited and learned knowledge that individuals are sustainably connected in a habitat where independence is rooted in the shared land and not in the possession of things. These conclusions are supported by the academic research discussed and analyzed in light of competing views.

While objections will be addressed throughout the book, I need to acknowledge philosophers like Nicolas Delon (2019) and Zipporah Weisberg (2019) whom I respect but who might question the claim of personhood status for great apes because of their human likeness over other species. This claim, to them, appears anthropocentric, sets up a dangerous hierarchy, ignores species diversity, and neglects widespread animal suffering at the hands of human factory farmers. I use apes as one example for the sake of simplicity and assume that all animal life is recognized as valuable in itself. Besides, it's not so much that apes are "like us"; rather, humans are, at base, like many primates and mammals in how feelings and emotions register and are expressed. My emphasis, in fact, is on ape/human differences in terms of the land ethic. Seasoned ape researchers like Wouter Wolf and Michael Tomasello (2019) point out that in comparison to humans, apes indeed are "a special case" among all animals by virtue of our shared phylogeny. As will be touched on in upcoming chapters, many human behaviors, even the "rational" ones, derive from an emotional foundation shared across species.

GENERAL OBJECTIONS TO THE ARGUMENT AND CLAIMS OF THIS BOOK

Not surprisingly, many people would object to granting personhood to great apes. Should the objections be enumerated and addressed? Many will be. I trust that the quintessential claims made in my book will adequately answer objections. Suffice it to say that from the beginning readers will see, for instance, how courts object to arguments for ape moral individualism in their reluctance to give away an inch of human dominion over what is legally considered property—animals. A related element of this book is not to dehumanize *Homo sapiens* by drawing parallels between them and great apes but, rather, to demonstrate how great apes are persons of their own accord, intellectually, socially, and morally in their tropical forest habitats. Does this not buy into a hierarchy, animal rights activists might object, by placing great apes atop an animal pyramid? There is no hierarchy per se; there are only varying degrees of difference between species. Besides, that's an unfair accusation since my text, by examining ape species comparatively, attempts a shift in cultural attitudes about the treatment of "nonhuman" forms of life.

Philosopher John Hadley (2015) says there are three principles of animal rights, namely:

1. Sentience
2. Equality
3. Evolutionary connectedness

While most scholars relate or combine these, Hadley claims they are not compatible and should be separated. Is an ape's psychology only hedonistic and not rational? In answer, is human psychology only rational without a heavy dose of hedonism? No one is saying ape-human psychology is equivalent, and in fact the differences, especially regarding eco-engineering (where apes seem to surpass many human societies) will be highlighted in my book. Is the argument of *An Ape Ethic* to place higher value on moral complexity and thus exclude non-ape species from consideration? Not necessarily, but that's not even the purview of this book, which argues specifically for great apes as self-governing persons of the tropical forests because they are eco-engineers. In the natural world there is no hierarchy of human "values" differentiating species across a threshold, only a struggle for existence with some mutualism by all plants, organisms, and animals.

Julie Smith and Robert Mitchell (2012b) are right to label a group of animal researchers as neo-Cartesian. Take, for example, Gary Steiner (2012) who, while granting wide latitude in ape conceptual thinking, makes some serious objections. Steiner refuses to grant moral agency to animals and says that it is only hypothetical. He says that animals

1. cannot position their individuality among others, especially not in the future;
2. cannot categorize their needs and desires in comparison to another's;
3. cannot conceptually manipulate reality regarding future events;
4. cannot think or communicate linguistically.

In a similar vein, Alain Morin (2012) claims that animals are unaware of their emotions, do not experience anxiety, and have only "primal self-regulation" without theory of mind. These objections will be addressed throughout my writing. Briefly, consider how apes are more than capable deceivers, which depends on imagining and anticipating another's beliefs, planning, and self-reflection. Those qualities are morally tinged. Furthermore, apes engage in emotionally charged reconciliation behaviors and alliances (including conspecific avoidance), pointing to their predictive thinking. Communication occurs regularly via gestures and vocalizations. They hide intentions and emotions from others for self-gain, and yet they sensitively care for others, cooperate, and even share food. Those actions are colored with good/bad behaviors, right/wrong understanding, or the biology of morality. Finally, if apes hold others responsible for their actions, they understand and act as thinking moral agents. My point, though, is how advanced ape cognitive and social behaviors enable them to interact as effective ecosystem engineers, in ways that many humans are not.

Paola Cavalieri (2015), one of the progenitors of the Great Ape Project along with Peter Singer, efficiently dismantles key objections. For example, the opposition might say the argument of this book derives values from facts in ascribing moral intentions to apes based solely on behavioral observation. Cavalieri says such critics are defending the status quo of human superiority and of human values. In turn, Cavalieri points out, such defense of human-only standards is a form of speciesism that rationalizes human actions against others. Reflect on, for instance, the human subjugation of horses over many centuries, even in times of war. The information and observations of *An Ape Ethic* show how great apes exhibit social duties in maintaining tropical forests and are, therefore, entitled to rights over those lands.

Certainly a serious objection might come in this form. To claim great ape personhood on, in part, the basis of how humans with mental defects are persons is an attack on humanity and a dehumanization of the individual so disabled. Cavalieri (2015), in this vein, notes how charges of eugenics were hurled at the Great Ape Project. The rebuttal is not so difficult, for one can see how those shouting dehumanization are once again making an appeal for species membership. Worse, some would counter and say that in leveling the playing field between humans and great apes one is guilty of species bias by excluding, for example, lesser apes, monkeys, and so on. Here, too, in answering, the counter claim is simple. Whether people want to believe it or not, great apes really are like humans (or, more precisely, humans are like great apes), and that is why so much emphasis is placed on them, as a first step, in "animal" personhood debates. Yet this puts aside my repeated emphasis on how apes differ from humans as more efficient ecosystem engineers.

As Cavalieri (2015) so eloquently says, a project that seeks to grant personhood to great apes is committed to social reform. For those who object to any variety of higher consciousness in apes, the data and arguments laid out in this book reveal not only an advanced bodily consciousness among apes but also innovation in their ecological consciousness as a land ethic. Personhood is an issue about persons. Imagine the enormous cultural shift that would occur by granting great apes rights over their own domains. That supposes that many so-called advanced societies would have to recognize the rights of all those people they so conveniently marginalize: the poor, women and children, undocumented immigrants, and disenfranchised minorities.

OF EMOTIONS, MOTIVES, AND ECOLOGY

Emotions are often involuntary expressions. Can we say the same about ecosystem engineering, where there is no real conscious intent per se but feelings, thoughts, and movements that aggregate into a result? Primatologist Frans de Waal (2019) says how blended emotions show that the ape body, as much as the mind, is central to ape existence. Since humans have emotional responses to nature, consider, then, how that sensitivity is magnified in apes, for example, the satisfaction of spotting, moving toward, and eating fresh fruit from a tree. An emotion is a mental state that affects physiology, de Waal might say, so it's difficult to separate those elements from ape contact with their physical world. Just as modern humans, mostly estranged from nature, are attached to their possessions and homes, natural elements and visceral relationships are the substance of and integral ingredients to an ape's home (habitat).

I'm suggesting that apes have an empathic connection to nature, where their minds and bodies fuse with the music of their sphere, much as a mother tree acts as a hub to supply nutrients to surrounding younger trees. Tropical forests, especially rainforests, offer a range of ecological niches for organisms to inhabit without competing with each other, leading to diversity. This level of variety is such that in some areas multiple primate species thrive together. But for some savanna chimpanzees, forest trees are part of the ape ethic since there is routine, intensely mutual, reciprocal interaction with wide scale ramifications for plants and other species. Eco-engineering among apes might not be conscious but seems to involve some version of care. This is perhaps too speculative, but the evolutionary history of ape species indicates some grain of truth about sympathy between apes and trees. Apes are abundantly capable of empathy. My educated guess, based on the empirical evidence I've gathered, is that the evolution of this *feeling into*, stemming from the intense mother-infant bond, carries over to ape attentiveness for the well-being of habitat, a synchronized, symbiotic recognition of mutual inclusiveness.

Do apes know how nature works? I think so, and by analogy offer this story from de Waal (2019). At the San Diego Zoo, the bonobo enclosure once had a wet moat, which was periodically drained and cleaned. Afterward, the moat was filled with fresh water. Once during this procedure as the caretakers left the moat area and went to the pump room they were distracted by bonobo Kakowet, an adult male, gesticulating wildly toward them. As it turns out, he was telling them to turn off the water since infant bonobos had slid into the moat. He "knew" they were about to be drowned. Consider, too, how simian forest sympathy is supported by behavior from lab reared apes who, when retired to sanctuaries, don't at first know how to react to grass or trees. De Waal (2019) suggests that emotion is to motivation as outcome is to function. In other words, behavior is grounded in motivation and function. While we can see this causal relationship in social situations of apes, I'm suggesting it also exists in their connection to habitat. This means that apes act in ways, individually and socially, that are emotionally and rationally environmental.

Let me propose some brief comments that draw from philosopher Mark Rowlands (2012), *Can Animals Be Moral?* Since there is a cumulative effect to my book about the environmentally driven moral individualism of apes, I'd like readers to keep some of the following thoughts in mind over the course of the next few chapters.

In his book, Rowlands (2012) indicates how animals evince moral emotions such as care, solicitude, toleration, and patience (his list). Like other thinkers, Rowlands uses the word "animal" and so lumps together many species. He also spends considerable time discussing humans, hypothetical humans, and equations. However, if I could impose on my readers, I'd suggest that whereas moral emotions apply to kin and others, in the case of apes they seem to relate to care for their environs. Whereas thinkers like Rowlands tackle big philosophical questions without directly using empirical studies, I try to employ the work of scientists while addressing theoretical and philosophical questions. As readers will see, especially in chapter 2 (dealing with cognition and intelligence) and chapter 3 (dealing with social behaviors), my book is textured with the work of many empirical studies aiming to lend credence to my notion of apes as persons expressing a land ethic.

For me, it's difficult to talk about an ape ethic without weaving into the discussion moral individualism and ecosystem engineering. The ape ethic, after all, centers on the "moral" behavior of apes to their land, a net effect of individual and group behavior. Beavers are eco-engineers, like humans. But in comparison with apes ranging in large forests, consider the difference of scale, both creative and destructive. Concern for another's welfare, says Rowlands (2012), is moral. In my case, I can't prove conscious intent in ape eco-engineering, yet in chapter 3 and elsewhere I show how ape intentions manifest themselves in social and physical environments. I'm at least hopeful that the questions I raise in this book will prompt not just concern but action by humans on behalf of apes, animals, and their biomes.

We can't equate human morality exactly with what apes exhibit because of different evolutionary paths and habitats, where many humans occupy, for the most part, artificial, constructed homes. For apes the setting is a tropical forest, and there seems to be an element of inclusive fitness in their dealings with place. Just as humans or beavers "care" for their habitats since the environs promote genetic fitness, so too apes but on a larger scale. Great apes are motivated to eat but evolved to live in balance with forests. Rowlands (2012, p. 14) says that contrary to the philosophical consensus of denying moral emotions to animals, nonhuman beings are motivated by emotions and considerations that have "moral content." While we can see this attitude in terms of mother-infant, kin, and other social relationships, I'd like to extend this moral reasoning in apes by suggesting they have no strong predatory instinct to ravish their environments. Of course this is true of other species, from beavers to giraffes. But since ape mentality and moral reasoning are so close to that found in humans, I'm wondering aloud if they indeed have a conscious land ethic.

A key difference here, between apes and humans, and which appears to bolster my point, is how the greatest eco-engineers, humans, have leveled wholesale destruction on plains, forests, and oceans. I suppose my question is, in terms of moral individualism relevant to the land ethic of eco-engineering, which species engages in the most productive gardening for itself and, as a byproduct, for others? Orangutan "care" for rainforests promotes the health and well-being of local and distant species, since these forests are integral to global air health which, in turn, impacts ocean temperatures. I use the words "moral" and "morality" to indicate, similar to Charles Darwin (1871, chapter 4), how social sentiments like sympathy and compassion appear in animals. But I stretch this idea, granted on little evidence, to ape concern for ecosystems. Their land ethic evinces an inclusive fitness factor with derivative benefits for others. Apes are not "moral" as humans; *Homo sapiens* evolved in different social and environmental circumstances. Rowlands (2012) suggests, and I insist, that human morality is not the benchmark.

Likely there's a genetic basis for eco-engineering. In addition to the obvious fact of manipulating tree branches and foliage for nests, protection, and food, those apes who have massively engineered their environs sustainably in mutual cooperation with others have survived. While tolerant reciprocity as a capacity is critical, so is the behavior. What I mean is that if we were to videotape them, apes clearly seem to behave as ecosystem engineers. A charge against me here is whether or not apes are capable of destroying their habitats. This might be unfairly exaggerated since ape genes that spread involve cohabitation with natural elements. Human genes have spread, it seems, more widely though not exclusively outside of natural environments. Who, then, among species with high intelligence, are the true collaborators with nature evincing toleration and reciprocity? Of course across species there are various forms of concern, but I suppose my point is that apes, more deliberately than others, seem to show a mutualistic care for their local and extended habitats in greater degree than many other species. I use thinking like this to call for ape dominion over their lands as self-sufficient persons of forests.

Arctic foxes, like beavers, are certified eco-engineers, but the scope of their positive impact is defined locally. Worms are ecosystem engineers, but many of them, and therefore their sustainable efforts, have literally been paved over by humans. By contrast, ape ecosystems are large and include thousands of square miles of trees that positively impact global health. Do apes "know" what they are doing? I can't definitively prove apes as conscious eco-engineers, yet given that apes possess beliefs about self, caring, and nurturing, and what constitutes a belief is really connected to a series of other beliefs, apes' intentional states concerning their environs could be considered a land ethic. Some critics of my idea would hesitate to attribute a land ethic belief to apes. Given ape/human similarity of deep history, the basis of a web of beliefs seems to be there. At the same time, given ape/human differences in evolutionary trajectory, the ape land ethic is not human. The forest is not an object of mental representation to an ape but, rather, a mode of being intimately attached to the vast, undulating physical sea of trees. This is why ape/human similarities are important for my discussion, but why the ape/human difference in how each species treats a biome proves to be a critical factor in my argument.

Rowlands (2012) says he can't defend moral agency in animals because it's too closely tied to responsibly approving or disapproving. He does, however, say animals are moral subjects who have emotional reasons for their motives. Great apes do indeed express approval/disapproval by holding wrongdoers to account for social transgressions or theft. The real question is how any "animal" as a so-called moral agent will act to another species. For me, the crux is seeing apes as moral agents to their environs. It's well documented (see, for example, Fouts, Patterson, Miles) that apes can act with moral agency toward humans. Some might claim in disagreement that great apes as moral agents to humans have been encultured. Indeed, human and ape children are likewise taught and learn how to act with socially acceptable behavior to kin and others. Are human children reared outdoors with instruction about habitat care? If an aggressive chimpanzee did not engage in reconciliation behavior, then he'd not know how to evaluate his behavior. But apes do engage in grooming and conciliatory behaviors. If they "think" about these types of socially motivated ("moral") activities, do they not think about their bodies engaged in their habitats?

Rowlands (2012) separates motivation *for* action from evaluation *of* action and suggests controlling one's motives makes for normative behavior. Readers will see, though, in the chapters that follow how apes express control of and thinking over emotions and actions. David Premack and Ann Premack (1994) show how chimpanzee Sarah applied inferential reasoning to obtain fruit. Autonomy is how and when one consciously decides on and controls her *ends*. According to Rowlands, this is not so in animals but only in humans. Rowlands seems to suggest, following others like Christine Korsgaard (to be discussed), that rather than considering motives and reflecting on actions "animals" react viscerally, though they might have a "reason" to act. This attitude seems to ignore the complex psychology of great apes who know each other (and even each other's history) in a social hierarchy. Rowlands says that without practical wisdom animals can possess only natural and not truly moral virtue. Based on data I provide, I will show how great apes are not always like small children prone to errors of social judgment, such as jealousy. Rowlands is not so sure one needs access to moral deliberations, though he goes on to say only some humans act on moral reasoning. This means that most animals share a human tendency of unreasoned action. His thrust seems to be that animals are moral in the sense that they exhibit and act on reasons of care for others.

This last statement is certain, as I will show. Moreover, I amplify this thought to the moral question of great ape ethical response to their environs. In stark contrast to most human societies, the ape "social" *mores* take place in a natural (not artificially constructed) environment. The physical habitat for apes is elemental to their social world and why I see them as acting toward it with care. Are apes at the mercy of motivations or can they think about and so control what motivates them? I believe this will be answered affirmatively in the chapters that follow. I spend time considering the moral individualism of apes and their ability to be ethical because if they can be moral socially and are, therefore, aware of their motives to act with consequences, there might be a parallel as to why they have evolved to be large-scale eco-engineers and not just suckers or gropers. I explore the problem of how they understand the relationship between the need to eat and the preservation of their food source.

THE WHOLE ORGANISM RESPONSE

I'm borrowing the heading of this section from Rebekka Hufendiek (2018) and her book *Embodied Emotions*, though I might stray from her focus. Emotions are shaped by culture as much as by evolution, and for my argument, the culture has to do with the ape land ethic. Theirs is a culture of earth and trees, consuming soil and leaves. Emotions can possess both representational and normative aspects, which, as Hufendiek suggests, means approval of rational norms is not necessarily similar to responses of emotional norms. While highly intelligent, as I show in chapter 2, apes fit into their habitats in a feeling way. There's a difference between emotions and judgments, minimizing a cognitivist approach, since ape bodies can almost spontaneously express emotions like pride, jealously, and shame. John Ruskin asked about the cathedral worker, *Is he happy at his task?* I ask, indeed perhaps naively assume, the contentedness of the ape in her forest home. Emotions and feelings can be like judgments since they are concerned with one's welfare. If what we call emotion is inadequate for humans, so the word is even more inadequate, in my accounting, for apes since any given emotion is interfused not only with the body/mind among others (like humans) but with the environs (unlike humans), seen, for example, in apes at rest, at play, at feeding in their habitats. There's a fluid continuum of perception, attention, feeling, thinking, and action in the soil or vegetation for apes more than separable components. How do we grasp that? How do I explain that?

The environment for an ape, following Hufendiek (2018), is not a bundle of disconnected elements separate from the organism. In fact, I'd say to describe an ape's habitat is to characterize, or give ethos to, the ape. The environment, suggests Hufendiek, has direct relational value to the organism, for example, as a home and food source. The ape body is not only the individual's but the habitat itself, where there is little difference between skin, bones, leaves, and branches, a collated unity of sorts so the feelings, emotions, and cognition are embodied not just in the ape herself but by extension in the pattern of the ecosystem. The ape land ethic is the ape's physical and mental ethos, nearly as one. That's a hard concept for us to imagine, I realize, since we tend to intellectualize everything. The environment itself can be filled with "normative" values, for example, go here, avoid there, approve/disapprove this action. Apes are often aligned to humans by virtue of their social behaviors, and I will discuss much of that in great detail in chapter 3; but my argument looks more to differences where the ape's world is really its garden domicile.

There's a valid question about how animal culture is human induced. Thibaud Gruber et al. (2019) say that human agriculture and development put stress on primate habitats to such an extent, in some cases, that food sourcing is altered. Because of their slow life histories primates don't necessarily adapt genetically but through behavioral flexibility. For example, in Uganda, chimpanzees will enact night raids on human crops. In reaction to deforestation in Indonesia, orangutans disperse into palm oil plantations to feed and nest. These are examples revealing forced modification of behavior demonstrating how some human activity deprives apes of their true eco-engineering capacities. Primates need access to dense forestation for feeding, socialization, and nesting. Furthermore, their feeding has a positive consequence of maintaining forest health for themselves and, similarly, nearby local people who rely on the medicinal benefits of certain leaves and plants (Lambert, 1998). Typically, chimpanzees and bonobos, our closest genetic relatives, don't raid crops but, rather, seek to extract nutrients and minerals from their forest habitats naturally and in accordance with inherited traditions or innovations (Hohmann et al., 2019). The ape diet has enabled, and in turn feeds into, larger brains and increased intelligence (Hopkins et al., 2018).

Visual perception is ecological and grounded in phenomena and discovery. While at times I might emphasize the brain, my overall thematic weight is actually less about the evolution of an organ and more to the environment of evolutionary adaptation of the organs that comprise the organism. As Sabrina Golonka and Andrew Wilson (2012) might say, what's critical is the environment that affords stimulation and action from perception. Apes are not stimulated in cages or zoo enclosures. Body and mind are constitutive in a natural habitat. The mind is neither isolated nor totally privileged, for the body is involved in cognition (Leitan & Chaffey). This is different from having subjective meaning in phenomenology but can be integrated into a naturalistic approach using ecological psychology, which sees cognition and emotion emerge from the organism adapted to its evolved environs.

Borrowing, in part, from Claire Michaels and Zsolt Palatinus (2014), here are some ideas from eco-psychology I will attempt to cover and apply to great apes as moral individuals because of their ecosystem engineering.

1. The ape and environment are inseparable, meaning that mutualism succeeds over dualism.
2. Information is part of a time-space pattern related to the ape in her environment so that data perception consists of the laws of physics and natural contexts.
3. Perception, attention, and emotion are evolutionarily embodied and deeply connected.
4. Detection is intuitive and not necessarily computational.
5. Perception services action through exploration.
6. Psychology is physical.
7. Physical perception (i.e., attention and memory) can, often, be more important than mental representation in temporal experience.

The nature and function of these interrelated components encompass the whole person, as in gestalt, giving rise to solutions through environmental circumstances. As botanist Verne Grant (1963) intimates, animals are not mechanistic but consist of the totality of cells and chemicals integrated into a coordinated system generating organization from disorder. Virtually the entire surface of the earth is inhabited by living organisms, from lichens and algae to insects and primates. Each of these species has evolved its special adaptations to survive and reproduce in a specific habitat, some narrow and some quite diverse. Perhaps stretching the findings of some a bit far (Wouter & Tomasello, 2019), I'd suggest that the emotional and mental states of apes, as in humans, can be shared when directed toward a similar object. By this I mean that synchronous connection is apparent for apes regarding their physical movements in and close connection to their tropical forest environs. Ape fitness for survival and reproduction is included in the forest. The bonds between ape individuals and among groups might be fostered by environmental interactions as much as social factors. The evolved psychological mechanism in humans to create social bonds from shared visual attention seems, in apes, to have grown from and into the context of their ecosystem.

Ecology is part of natural selection. Although mine is not a technical book about bio-ecology, it goes without saying that some ideas from G. Gaylord Simpson (1965) on the geography of evolution and Eric Pianka (1974) on evolutionary ecology are evident. Bio-geography is concerned with distribution patterns among plants and animals based on mutual and competitive interactions. Individual self-preservation for reproductive success and hence survival of the species comes via mechanisms of metabolism, behavior, homeostasis, ranging, and population density (Calder, 1984). In this way, niche construction involves coping with and conforming to an environment, from organisms dormant in dry soil and old seeds waiting for rain to hibernating bears, Pianka could say. Tropical forests participate in community ecology or variously assembled niches, since ecological systems are open, unlike the plastic confinements of most modern humans. Apes, conscious of and adhering to spatial restrictions in bio-geography, are adaptable ecological engineers shaping and shaped by their habitats in terms of how they utilize time, space, matter, energy, and birth intervals to become inclusively fit with their ecosystems.

DOMAIN FOR APES

Following political philosophers like Sue Donaldson and Will Kymlicka (2011) and Eva Meijer (2019) I see human and animal interspecies communities, with limitations. For the sake of my argument, human-animal community is minimized to elevate the idea of great apes as community communicators within their own *Umwelt*. Some thinkers will criticize this notion as fixing apes to one locale. True, but that is my point. Where much human business has stained the planet, ape activity, rather, has helped tropical forests bloom and green. It's a question of how a group organizes its environment. Besides, as you will read, I often note that humans are, in fact, phenomenologically connected to apes and other species through the notion of Gaia. The differences between apes and humans, I assert, are not so much about separation because we share the same biosphere. The difference is about concerns. Although apes eat much over a range that does not mean they are ravaging their forests. Instead, it implies they are seed distributors and plant cultivators on a widespread scale. Beyond any interspecies democracy Donaldson, Kymlicka, and Meijer call for, apes are persons in their own right who have earned, by virtue of their history, forest sovereignty.

Donaldson and Kymlicka (2011) announce a type of human citizenship to animals. They argue that welfare claims (where humans grant rights to animals) or ecology claims (valuing ecosystems over animal rights) are misled causes that elevate human needs over animal interests. I agree. But note how I say that great apes are moral individuals because they engage a land ethic as ecosystem engineers. In other words, evident from their evolutionary history, apes and other nonhuman beings continue to ably express their own interests, which are often very different from and in conflict with human desires.

Donaldson and Kymlicka (2011, p. 19) argue for a fundamental change in human political thought that grants "inviolable rights" to all animals. I agree. Where we might seem to differ is in my use of the word "personhood." These authors say that granting personhood to one species (e.g., humans) diminishes all others. Perhaps that is true, depending on how the word "personhood" is used. These writers are political scientists, so their conception of person is more to the legal and less to the biological. I am not trying to legislate who is or who is not a person. I use the word in a figurative sense so that readers can grasp the depth and range in the moral and ecological character of apes. The expression of *person* has more clout than the face value of the word "animal." Besides, I qualify the word "person" so: forest person. As anyone can see from the title of my book, I admit there's a question of ape personhood. In terms of the truly nonpolluting and sustainable ecosystem engineers, humans are inferior to apes, but we are, nonetheless, apes as well as persons. In line with the logic of Donaldson and Kymlicka, is personhood a "barrier" (30) to rights if humans do not employ the "person" capacity of forest sustainability like apes?

Hardly, of course, and that's my point. All I'm saying is that it's easier to imagine apes as eco-engineers on a grand scale (compared with Arctic foxes or beavers) when we figuratively see them as forest persons and not as "animals" or instinct-driven machines. I use the word "person," therefore, in a much more metaphorical sense than others. Because of how they evolved, apes have an express and privileged right to their domains that humans do not have to that land. Apes are forest persons as most modern humans are village, town, or city persons. I don't see in these linguistic qualifications any obscurity, as Donaldson and Kymlicka (2011) might object.

The biological facts speak for themselves, as I will demonstrate especially in chapter 2 on intelligence and chapter 3 on social behaviors. So far, you see my emphasis is on the moral individualism of great apes as ecological engineers which awards them dominion over forest habitats. Apes evolved to remain in and care for tropical ecosystems. That fact lends credence to my claim of their sovereignty, and I ask that readers hold to that idea while reading this book since all the other parts revolve around it. The upcoming chapter will elaborate some generalities covered in this Introduction and offer new insights; then, lots of details about ape cognition and moral behaviors related to my argument and claims will be covered in later chapters.

It's important, therefore, to keep my basic argument in mind as I delineate related claims in the many pages that follow: Great apes as morally individualistic forest persons exercising a healthy land ethic as efficient ecosystem engineers should be allowed sovereign dominion over tropical and rainforest environments.

Chapter 1

The Case for an Ape Ethic

This chapter provides the theoretical and philosophical groundwork of my book's subjects and claims. I expand on key ideas of moral individualism, ecosystem engineering, and the land ethic outlined in the "Why This Book" section of my Introduction. While there is some detail here, these topics will be covered more fully in chapters 2 and 3 that feature a critical examination of academic research and ideas on primate intelligence and social behavior, particularly as I relate those to the ape ethic of eco-engineering.

What is it like to be a great ape? I hope to answer a question like this. I also hope to go beyond the effort to claim legal status for great apes and offer the controversial idea of moral individualism to apes since they have their own land ethic evidenced by how they are sustainable and sociable ecosystem engineers. If anything, this highlights the differences and not the similarities between great apes and human beings.

What is it like for an orangutan to be human? That question is as fair as the one famously put forth by philosopher Thomas Nagel (1979), *What is it like to be a bat?* (to be discussed). Human morality, as biologist Richard Dawkins (1993) says, is based on discontinuity. I suppose my question about personhood is why humans hold one ethical code for themselves and another, or more precisely, none, for "animals." What about human morality when a great ape behaves intellectually and socially like a human, as shown by Laurent Dubreuil and Sue Savage-Rumbaugh (2019)? Such morality cannot then be arbitrary.

The damage inflicted on orangutan habitats by corporate plantations and logging companies has a dramatically negative effect on the psychology of this great ape. In order to illegally obtain a baby orangutan for trafficking, the mother is killed and the infant is abducted. Human actions like these amount to persecution. Gisela Kaplan and Lesley Rogers (2000) tell the story of ex-captive orangutan Judy. This orangutan person was virtually hairless and raised by people to be human, in how she routinely walked bipedally. Judy did not care to be reintroduced to the forest. Judy most likely saw her mother slaughtered while she was sold to the highest bidder like a piece of property. On her last attempt to be a wild orangutan, Judy actually walked out of the forest along a footpath to the main road, sat on a bench at the bus stop, boarded the bus, and returned to the Sepilok Rehabilitation Centre. What a tragic reversal of nature's course for this ape's ethic of moral individualism. Great apes evolved and adapted to their homes where they are comfortable. They can also feel pain, suffer, grieve, and experience depression, mostly because of environmental and psychological challenges thrust upon them by some humans.

In the pages that follow the continuity between great apes and humans will be explored. Differences are equally important. As Sara Waller (2012) points out, any focus on finding similarities blocks an objective ability to understand the true nature of animal cognition while typically placing humans at the center of the story. The decisive, and not superficial, elements of ape personhood are covered. This is not to say that by virtue of their personhood great apes should be integrated into human society. Rather, the aim here is to help people recognize apes as having rights, duties, and responsibilities on their own turf. They should be left to their own devices in their own environs. These arguments have been made before, but there is great urgency now for several reasons. Tropical forest habitats, already eroded by human intrusion, are in peril. Rainforests are vital to earthly life, acting like breathing bellows for the world's biosphere. Just as humans care for human life, they should value the lives of species like great apes who are remarkably similar to them.

All life forms, for that matter, except perhaps certain bacteria, viruses, and cell mutations capable of wide scale destruction, should be respected. This is because, as demonstrated in the Gaia hypothesis by James Lovelock (1979), all manner of species, whether microbes, insects, plants, and even some parasites function in symphonic unison to regulate and maintain earth's vast oceanic, geographic, and atmospheric resources to promote and sustain all varieties of life forms.

One can only grasp what it means to be human by understanding that which is labeled as nonhuman. This is a moral issue since there is no strict divide between human and nonhuman forms, speaking primarily of *Homo sapiens* and extant great apes (orangutans, gorillas, chimpanzees, and bonobos). Ethnoprimatologists like Nicholas Malone and Erin Riley go as far as saying that there is no real distinction between what is called human and what is termed nature. There are, rather, "multispecies entanglements" (Malone et al., 2014) and "ecological and cultural interconnections" (Riley, 2006) shared among humans and primates where, for example, human farming could overlap with primate resource ranging. Social and ecological cohabitation between humans and primates is growing on an increasing scale and, might in part, demonstrate species correspondence.

But modern humans are far different agriculturalists and pastoralists than great apes, since people use machines and chemicals to deplete resources whereas apes use their hands and feet to cultivate nature. This critical difference in land use raises the question of how humans and apes could cohabit some areas. I don't think they can, and why apes should be granted forest sovereignty. Besides, local and corporate farmers in, for example, Indonesia, classify orangutans as pests and treat them as such.

The likeness among great ape species, including humans, is remarkably close, with differences more in degree than kind. Of course there are enormous leaps in those degrees. Nevertheless, philosopher Carlo Alvaro (2017, 2019) intimates that many animals possess, broadly speaking, moral characteristics. Great apes express both moral and ethical behavior. Generally, one can define morals as outward social norms of conduct, whereas ethics is more universally concerned with an individual's perception of right/wrong behavior. Individual apes act socially in a group, but each also has her own personal temperament. The terms are closely related and used in proximity hereafter, but ethics and ethos are mostly related to one's character. One must approach a definition of human by exploring and experiencing great ape forms and their social attitudes and interactions with humans, as done by Laurent Dubreuil and Sue Savage-Rumbaugh (2019). Humans need to acknowledge how they evolved in fits-and-starts from a number of early hominin species (i.e., the early ancestors of modern humans) through the mixing and matching of traits.

What is a primate, a category that includes great apes? According to biologist Robert Martin, as summarized by Robert Boyd and Joan Silk (2009), traits of most primates include, for example, gripping ability and fingertip sensitivity, reliance on vision and depth perception, a long pregnancy and late maturation, a large brain and behavioral flexibility. Consider how many traits on this list are held by humans and shared with large primates ranging from baboons to mountain gorillas, to say nothing of joint intelligence and social behaviors.

How does one define what is human? Does one study only customs and behaviors from recorded history to the present? Which disciplines bear on crafting any definition of what is human? Certainly the arts, philosophy, and psychology weigh on how one understands what is human. How can evolutionary biology also help us understand, in a broader sense, what it means to be human? If there are discrepancies between philosophy and biology in defining what is human, how can they be reconciled? Biologists assert, following Charles Darwin (1859), the continuities between animal and human forms. Taking that perspective into consideration, one can concretely say that across mammalian species there is a mother-infant bond, parental care, altruism, cooperation, and empathy. Moral decisions are emotionally based in perceiving and knowing how another feels, and especially in reacting on that inner feeling. In great apes such as orangutans, gorillas, chimpanzees, and bonobos, one can go even further and say that they possess cognitive functions, intelligence, and social behaviors very close to what is human.

On fundamental levels, then, what is called human is in fact shared with many animals, primarily mammals, especially with nonhuman primates, and very closely with great apes. What Frans de Waal (1999) calls anthropodenial, or how humans deny their own animal nature, was first addressed by Tom Regan in 1983 (2004) as human chauvinism, when one denies that another species possesses the very characteristics of the chauvinist. Humans supposedly make progress, exert will, have government, and enjoy happiness. Apes do, too, evident in an important book by Frans de Waal (1982) called *Chimpanzee Politics*. More than investigating the definition of what constitutes being human, I am asking, really, what is a person? Can a great ape be considered a person? Why is a human fetus, which has not experienced any cultural life, considered a person and yet a forty-year-old female orangutan who has lived in the wild and captivity among others and who has nurtured offspring not considered a person? Why are people who commit heinous crimes still considered persons, or worse, labeled as "animals"?

LEGAL RIGHTS AND PERSONHOOD

Animal personhood is a vast subject, legally and philosophically, so all of the commentaries cannot be covered in this section or indeed in this book. Philosopher Mark Rowlands (2019), for instance, tries to move the debate away from legal and moral considerations and tackles the definition of a person from a metaphysical angle. Sue Savage-Rumbaugh (2007) talks about the vocabulary of personhood and notes how different words and expressions deliberately separate human/nonhuman mentality and render some higher thinking "animals" as experimental subjects and not as social persons.

To answer those who question the usefulness of a personhood argument, Shawn Thompson (2019b) puts it succinctly by noting that existing animal cruelty laws simply don't go far enough. Ape personhood advocates seek a measure of equality saying that, in contrast, no human being would consider relinquishing his or her rights as a person. *An Ape Ethic* is not a legal argument but, nonetheless, a call for recognition of great apes as persons of tropical forests. Cognition, for instance, is not sufficient for personhood. What's at stake is how the cognition is used. My book attempts to demonstrate that ape cognition and intelligence are connected to eco-engineering (a land ethic) and to social behavior (a moral ethic). As moral individuals, apes are persons. Presented in this chapter are some salient ideas and points to consider over the course of reading the entire text.

In April of 2017 the Nonhuman Rights Project under the direction of lawyer Steven Wise petitioned to have *Black's Legal Dictionary*, an authority in the field, change its definition of person. In terms of legal theory, a person is any being the law recognizes as capable of rights or duties. The emphasis here is on *or*; previously, the definition used *and* instead of *or*. The Nonhuman Rights Project, for example, has filed for writs of *habeas corpus* (protection against illegal imprisonment) to argue that great apes are autonomous beings often held against their will. *Habeas corpus* literally means "that you have the body" and suggests, for someone whose liberty has been abrogated, produce that body for someone else who can intervene and assert rights. This legal concept probably goes back to Roman times and is stated more explicitly in the English Magna Carta of 1215. As it turns out, great apes are quite adept at recognizing possession, harm or remedy, acting with responsibility, and weighing duty to individuals or a group.

In her July 2015 opinion denying *habeas corpus*, on strict legal grounds, to two chimpanzees at a Stony Brook research lab, New York Supreme Court Judge Barbara Jaffe, nonetheless, acknowledges how chimpanzees demonstrate intelligence, communication abilities, self-awareness, empathy, emotions, compassion, play, and even humor. Many mammals, from wolves to elephants, express such traits, but degrees of some of these characteristics are indubitably heightened in great apes. Similarly, in a landmark ruling of 2014 in Buenos Aires, an orangutan named Sandra was granted the right to be represented as a nonhuman person in court. As is commonly known, the Balearic Islands in 2007 legislated personhood rights for apes. Judge Jaffe in New York concedes that legal personhood is not synonymous with being human. One can have "human" traits and characteristics and not have rights under the law. This attitude is a bit puzzling, for it should be self-evident that all beings have rights if they are generating some form of normative community relationship with others. For example, think of human prisoners, as with apes in zoos or confinement. Overall, many cultures do not respect most animal life, but that thought will be amended briefly elsewhere in this book. Anyway, this defeat prompted Wise (2017) to summon philosophers with animal cognition and behavior expertise to show how chimpanzees, and by extension all great apes, are responsible and have duties in the wild and in captivity: parenting, sibling care, policing, cooperation, mutual aid to non-kin, contractual promises, systems of chores, and, significantly, "moral behavior."

In a case involving cruelty to horses in India, judges there (Sharma, 2018) remarkably ruled that the government of India has a burden to protect the "moral values" of animals. Those justices recognize a person as any entity to whom rights or duties may be attributed. The Indian judges go as far as declaring personhood for the entire animal kingdom in their state of Uttarakhand where even stray or sick animals are to be henceforth considered citizens entitled to healthcare. These sensible judges cite a number of sources and ideas similar to what's amassed in my book. Moving beyond thinkers like Tom Regan and the Indian judges, philosopher Elizabeth Foreman (2015) says that even the notion of rights is anthropocentric; rather, she focuses on agent attitudes and not violations of rights. She says that bad treatment reflects a moral stance. Morality is not a matter of limiting action but evaluating actions, as I attempt. Of course, I am not opposed to rights for animals.

Here are some duties categorized by philosopher W. D. Ross in Richard Norman (1998, p. 98). As will become clear over the course of this book, many of these duties are equally applicable to humans or great apes in some form or another.

1. Fidelity, or fulfilling promises and debts
2. Reparation, or correcting an injury
3. Gratitude, or acknowledging a service
4. Justice, or fairness in sharing
5. Beneficence, or helping others
6. Self-improvement
7. Non-maliciousness

Number 6 can be questionable unless it is limited to physical health. Apes indeed self-medicate. Many apes have also demonstrated a high degree of curiosity in human productions like art making, in machines, and in mechanical processes. Legally, what is the object of duty? Is it utilitarian, the good for the greatest number? The problem here is that, as philosopher David Cooper (1998, p. 252) would say, any obligation with a concern for the greater good neglects the "personal character of duty." The ape ethic of moral individualism as an ecosystem engineer promotes duty for self and for others by carefully avoiding wholesale destruction of food resources and habitats. What is the purpose of duty, then? A question hovers around whether moral psychology fundamentally differs between great apes and humans. That psychology seems to differ in practice related to land use, as I hope readers see by now.

As New York Court of Appeals Judge Eugene Fahey (2018) says in his opinion denying the appeal of the Nonhuman Rights Project on behalf of imprisoned chimpanzees, great ape incarceration and restrictions on *habeas corpus* pose ethical dilemmas. He stops short there, though. While the court does not appear to assert the moral agency of incarcerated chimpanzees, it does recognize their moral rights and positively cites philosopher Tom Regan (2004) who pushes the case for animal rights. Leaving legal personhood to scholars like Wise, a distinguishing argumentative difference, however, is that in their natural habitats apes are persons who demonstrate rights and duties to self and others through the adapted stewardship of their Edenic gardens. They evince moral individualism with an ape land ethic. Taking a great ape out of his natural habitat does not eradicate but only cripples his ethic as a moral agent.

Each ape has a character, in the sense described by German philosopher Arthur Schopenhauer (1840). There's a fixed ethos or temperament influenced by environment or learning. Beyond mirror self-recognition, "languaged" and signing apes refer to themselves, that is, recognize themselves as distinct from others, and talk about themselves in place or time other than the present, even using signing to deceive others, implying knowledge of another's mental state. That's personhood, says philosopher Paola Cavalieri (2015, 2016). Yet many human beings doubt the personhood or inherent rights of other living beings, especially sapient and sentient ones. Sentience is the capacity to suffer mentally as well as physically. I hope in this and the chapters that follow to show, at a minimum, how great ape sentience and sapience are used to construct social and physical environments. Many animals are sharply aware of the pain inflicted on them. It's common knowledge that animals, and especially apes in psychologically human ways, languish when removed from their forest homes. As ethicist and theologian Andrew Linzey points out (2009), people might argue that it is permissible for animals to suffer because they don't reason or have language. That's not true. Would it, then, be more permissible for a normal human adult to suffer, in a situation where a choice had to be made between the human and an ape, since only the human could supposedly be reasoned with and spoken to? This hypothetical situation does not discount assertions that great apes reason and communicate with others. Animals may not demonstrate all the characteristics of human language, but they are effective communicators.

Rather than the question of whether or not they suffer, let's ask, whom do they harm? Some might say this question is irrelevant concerning rights, so let me rephrase it. Humans are hurting ape habitats and consequently causing them to suffer; meantime, apes sustain ecosystems that benefit many species. In general, orangutans and gorillas mostly avoid unprovoked, nonsexual aggression. Should an animal who harms a human be punished? This question was raised by preeminent primatologist Biruté Galdikas (1995b). One devilish orangutan, who was believed to have harmed conspecifics, was actually relocated far outside the boundaries of the field station. In an orangutan such unprovoked aggression is unusual. Typically, even if for adaptive reasons, chimpanzees exercise a high degree of self-control in spite of their aggressive tendencies relative to each other. In zoo confinement, chimpanzees have been known to throw rocks and feces at humans, but this could be out of rage for their imprisonment. Of all great apes, bonobos seem to be the least aggressive, which is not to say they don't engage in conflicts.

While punishment for apes who harm humans might seem germane to my discussion, it is not, simply because my argument calls for ape dominion over their own tropical forests with very little human intervention. What I'm saying is that apes demonstrate moral behavior within their groups and to their habitats. Apes don't *seem* morally relevant to us (but only to their companions and competitors), but that's not so, precisely since the empathic control they exhibit over their environs has a positive outcome relevant to atmospheric sustainability for species far away. In turn, a stable atmosphere helps regulate the oceans.

Meantime, a debate about the legal personhood of machines and robots is ablaze. This is not difficult to understand. Many humans tend to be an acquisitive and contentious lot, so the question of robot or artificial intelligence personhood boils down to liability—whom to blame if a self-driving car kills someone. Hartini Saripan et al. (2016) actually claim that robots of today have sentience. Yet by their definition, a robot is a "goal-achieving mechanism." At what costs to others is that goal achieved? If mechanical, how sentient; how does it suffer? In the famous trolley car scenario (stop the train by sacrificing one person to save five people) we know the robot would become utilitarian. What if the five people are environmental terrorists plotting devastation while the lone person killed holds the cure for global warming? Based on the evidence I will present in chapters 2 and 3, it seems strange that we'd assign human-like personhood to artificially selected robots but not to apes. There's a fallible, human maker behind a robot who is liable; on the other hand, natural selection is the maker of all organic life forms. That's a huge distinction. As Darwin (1859, chapter 1) notes, natural selection is successive whereas artificial selection is accumulative. As of now, machines are not designed to fill an ecological niche but to satisfy a capitalistic agenda that typically erodes environmental health. Robots are clones of human needs, desires, and vanity to maximize human rational potential. Great apes are naturally more than capable of making ethically rational and emotional decisions.

Historian Yuval Harari (2017) talks of, in his words, the second incarnation of *Homo sapiens*. He imagines bioengineered cyborgs as part flesh, part hardware, and part software. Nano computers will be installed in human bodies, so-called advancements which have derived from experiments harmful to many monkeys. Harari speculates that organic life forms will be replaced by inorganic matter where technology will "re-engineer human minds" (46), manufacturing *Homo sapiens* version 2.0. In near whimsical language Harari says that humans will enjoy an *upgrade* by merging their bodies with computer devices so as to shed their "animal" likeness. Where, in such a world, is personhood? It is precisely this line of thought, and I'm not sure if Harari advocates it or not, that has propelled humans to disenfranchise themselves from the natural world. Worse, super technologies in agriculture and mining have forced nonhuman life forms from their natural habitats. Rather than mechanically engineering a human version 2.0 we might want to focus on considerations of climate change that could be effected by altering existing attitudes.

Can a robot's so-called social skills discourage someone from turning it off? That's the question asked by Aike Horstmann et al. (2018), and perhaps not surprisingly the answer is yes. The study shows how people ascribe emotions and social norms to robots and actually perceive them as alive, especially if there's some assumed degree of intimacy between the person and the machine. The "voices" of robotic personal assistants are now programmed with "feeling." How can one view a robot, objecting to having its own power terminated, as autonomous and yet not see a great ape as autonomous? Apparently there is a media equation theory which says that people act socially when interacting with a functional machine. While the authors admit that a device cannot have its own thoughts and feelings, the sophistication of robots and computers is such that because of human evolutionary history people are naturally developed to respond to social cues as if the machine is animate, since robots can be programmed to express human-like emotions. Would some people enjoy an afternoon with their children at a petting zoo and then not feel any compunction eating baby animal flesh (e.g., veal) for dinner? Harari's human 2.0 could easily veer into a realm of compassionate dissonance.

Furthermore, the question about theory of mind capabilities in robots has been raised by Brian Scassellati (2002), whose paper begins with the disturbing assumption, "If we are to build human-like robots . . . ." Meanwhile no definitive conclusion is offered or inferences about such a scenario. Evolutionarily and developmentally humans and great apes have a theory of body model (which is mechanistic evaluation); above that is a theory of action model (to determine agency); and even more sophisticated is the theory of attitude model (tapping into another's mind). Theory of mind and mentalizing are intentional stances. Is a robot so capable? The problem is that the theorized humanoid robot has no evolved genotypic traits but acts on and learns by program code developed by a contemporary, technological geek. We know from evolutionary psychology that primate, ape, and human brains have innate modules that operate from birth, maximized and developed via social interactions, all naturally selected and replicating by virtue of survival success over millions of years. For instance, notice how a newborn responds to and even imitates the facial expression and gaze of its mother.

Let me take one example of how artificial intelligence cannot fully understand context. Without going into detail, consider the conveniently labeled, for marketing purposes, home personal assistant devices. These interactive robots have mistakenly interpreted commands or, worse, have inadvertently surveilled homeowners' conversations only to email that transcript to a third party. How are actions like those rational decisions? Readers, I'm sure, have been frustrated by AI phone operators who ask a question, and then, when you answer, the computer voice suggests something ridiculously erroneous. On the other hand, the two substantial research chapters that follow will thoroughly cover the emotional and rational intelligence of great apes as well as their demonstrated theory of mind. In any case, theory of mind need not be the sole criterion in determining the intelligence of another species. This book is not about artificial intelligence or robots but about the evolutionary connections between humans and great apes. Adding to my claim for ape personhood is how they, contrary to robots and humans, nurture themselves and their forest habitats simultaneously by mental as much as physical engagement. Perhaps robots will one day be able to employ an advanced theory of mind. Robotics enters the discussion only since it is disturbing that assertions of personhood and mentality for machines are foregrounded in the popular media while great apes, among other animals, and the plight of their habitats are mostly neglected but for hot "newsworthy" moments.

The question of legal status for other living creatures matters since humans dominate the planet and some people have eviscerated natural habitats. This is nothing new. Naturalist and explorer Alexander von Humboldt warned about deforestation in Brazil in the early nineteenth century. Perhaps limitations should be imposed on human population growth and not on apes' ability to thrive stably in their own ecosystems. Although Steven Wise (2000, 2002) cogently makes the case for legal rights, the attention here is for the scientific and especially the social bases for great ape moral individualism evidenced by behaviors around ecosystem engineering. Great apes, like humans, are neither merely "animals" nor machines. They are, like most humans, feeling and thinking persons who demonstrate care and concern for their own and another's welfare. Would personhood have to be limited to mammals, only to primates, or specifically to great apes? Is personhood status the only way some humans would not enslave primates and apes for their own desires or research needs, not destroy their forest homes? Would personhood status preclude humans from killing ape families so that infants could be sold as pets? Probably not, but it's a good start. Why are humans the most successful in having mastered all others when they know that intelligent species consciously express personal feelings and shared emotions? The question about what constitutes a person only raises more questions because it calls for a special category or preferred class. The majority of humans have granted themselves top rank and refuse to relent, but there are problems with the designation of human person.

Dare one ask, for instance, if human infants are persons? They do not have the range or depth of emotional intelligence seen in adults. Is someone in a catatonic or mentally impaired state a person? No one can truly measure or understand that person's internal narrative or thinking. Meantime, many researchers like Sue Savage-Rumbaugh (2007) have successfully comprehended parts of ape thinking and intentions by communicating with them. Given human technology, even if an inactive catatonic or a severely mentally impaired person's thought processes could be tallied, the outcome counts. Do random and incoherent vegetative brain waves measured by a machine quantify personhood more than a male orangutan consciously roaming and deliberately foraging in a rainforest tree canopy? Great apes and especially orangutans evince an ape ethic because the outcomes of their deliberations can be accounted for by their actions, which typically show consideration of rights or duties toward themselves, others, and, especially, to their environment.

Why can't people become familiarized with apes and not see them as zoo property or entertainment curiosities? Philosopher Eva Feder Kittay (2005) argues persuasively for moral personhood to humans with severe cognitive impairment. Is it easier for one to imagine himself as a great ape than as someone with advanced dementia? Perhaps that's an unfair question, but it's pertinent to this discussion. Just as someone with a serious mental deficiency has interests and is a person, so too is a normal functioning ape in the wild. Besides the definitions of personhood which are top down, consider an ape's demonstrated capacity and ability to apply personhood (individual consciousness) to others and to ascribe or take away another's interests. Apes, too, are moral agents and not simply moral subjects as many humans treat them. This means that included in moral individuality is moral personhood. Joshua Smith (2014) makes clear that apes, and especially the zoo-housed gorillas and orangutans he observed, differentiate humans based on familiarity and how well, or not, any human acted toward them. Orangutans seem more open to new relationships with strangers, even though they are semi-solitary in the wild. This makes sense because by their nature orangutans have a dispersed social system and, therefore, are open to persons at a distance.

Mark Rowlands (2019) makes a personhood argument based on the thinking of seventeenth-century philosopher John Locke. This viewpoint says a person is one in possession of, in this order: consciousness, rationality, self-awareness, and other-awareness. While he is familiar with the empirical evidence (I cite much more), Rowlands makes an argument even more speculative than mine. Plus, his notion of personhood is applied to all "animals," basically any moving being. Rowlands takes Locke's criteria and, blending them, argues for a metaphysical person. This is in contrast to how I use empirical studies and philosophical ideas to argue for a forest person, a real and knowable being who expresses agency toward her kin and environment. There's nothing metaphysical about that. Nonetheless, Rowlands is useful since he sees a person, human or animal, as one in possession of a unity of mental states, and not merely successive mental reactions. I'd argue that any continuum of one's mental conditions contributes to embodied awareness in ecosystem engineering, where emotional (e.g., happiness) and cognitive (e.g., foraging strategy) states are elicited by presence in and reaction to the habitat.

MORAL INDIVIDUALISM

Let me apply what philosopher James Rachels (1990) calls *moral individualism*, where individuals are treated according to their characteristics on a case-by-case basis. Individuality depends on autonomy, which in turn assumes a distinct personality. Apes are not an army of robots all making the same rote movements repeatedly. Perhaps Rachels is following the lead introduced by G. E. M. Anscombe (1958) that cleverly foisted virtue ethics (moral character rather than consequences, duties, or rules) over the Kantian focus on "humanity" and the utilitarian emphasis on numbers over individuals. With their high degree of intelligence, it's an open question as to whether or not an ape can decide what type of person he or she wants to be. But to be clear, moral individualism is a social practice. There are rights in the social order of apes because others can make claims against someone infringing on those rights, for example, an unprovoked transgression. Field observations from Goodall who studied chimpanzees (1971, 1986), Dian Fossey who studied mountain gorillas (1982), and Biruté Galdikas who studied orangutans (1995b) reveal, generally speaking, how there is a mutual respect for personal and group boundaries unless aggressively violated. Consider the case of the young gorilla, Harambe, who, regrettably, was shot to death by his caretakers in his enclosure at the Cincinnati Zoo in 2016 after a human child entered it. Outside of normal responses, Harambe did nothing to harm the boy who trespassed into his enclosure. Rather, the human caretakers under very stressful circumstances perhaps presumed that Harambe would inflict mortal danger on the boy. The gorilla was likely killed because of what others assumed he might do, not for what he did. Surely the caretakers grieve to this day, and maybe this incident gives us pause to ask why apes are incarcerated in zoos and not given free rein in the forests to which they've ecologically adapted. There are other such cases, notably from October 2007 when chimpanzee Johnny escaped from his Whipsnade Zoo enclosure in England and ran to a nearby meadow. He was then shot and killed. As with the case of Harambe, was there an investigation that included advocates speaking on behalf of great apes? The killing of great apes in situations like these is only presented as a moral decision to prevent harm to humans without concern for the ape.

Natural selection favors genes that foster altruistic behavior toward kin. This cooperative attitude has been sifted over evolutionary time so that one can act kindly to another as if he is kin. What humans call moral behavior can find its way back to altruism, reciprocity, trust, and empathy. Philosopher Colin Allen (2001) claims that empathy is the gold standard in determining moral behavior. All of these characteristics are exhibited in great apes and indeed among other mammals. I go as far as suggesting that there might be some inclusive fitness factor in how apes interact with their biomes. A basic moral code dictates that one should not harm another without good reason. Self-preservation is implicated here, for sure. Apes are more likely to follow this code of self-control than many humans. Certainly human morality might be associated with feeling good when doing what is right. Nevertheless, outside of altruistic or reciprocal behaviors, apes might simply be more restrained in their subjectivity and, therefore, cautious to act. Or they act in ways that are, let's face it, not maliciously human, although some people immediately judge them according to human-made moral criteria.

Morality is not a human invention, nor is it a human-only principle. As noted by Douglas Allchin (2009) and others (Trivers, 1971; Alexander, 1987; de Waal, 1996a), morality is an evolutionarily constructed behavior, and there are analogies in great apes. Certainly, Robert Mitchell (1993) is right to question whether apes have their own moral standards, partly the inquiry of this book (i.e., their land ethic). Nonetheless, the human moral circle is not narrowed by including the similar behaviors of apes. Rather, human morality becomes more comprehensible. Concerning morality, on which much has been written (Tague, 2016, chapter 2), origins stemming from mother-infant interactions go to the basics of a signaling system for conspecifics to distinguish between play and aggression. Subsequently, this basis set the tone for norms dealing with approval/disapproval and restraint. Breaking rules of play or social engagement, or ignoring signals of nonaggression, led to exclusion or even disempowerment. Natural selection endeavored to weed out consistently bad, costly behaviors.

As noted, ethics is to character and morality is to norms, drawing from the Greek and Latin etymologies of those words. By virtue of their distinct personalities and temperaments, apes exhibit character traits and, in a group of varied individuals, follow norms of behavior to accommodate the diversity of persons. The nitty-gritty, not the entirety of morality, is in feeling. How any sentient and sapient being acts on such feeling depends on her character. There is no social contract as per the seventeenth-century Royalist philosopher Thomas Hobbes, since one is not needed. Nevertheless, apes maintain nonaggressive relationships despite conflict, using negotiation, alliance, consolation, or even avoidance as a social "transaction" or contract, with some underlying genetic elements (e.g., mutualism), among peers and individuals to compromise (de Waal, 1996b). Apes manage their emotions, feelings, aggression, and economy in a self-regulatory manner. Welfare for others begins with self-care, and each of these concepts implies approval/disapproval of actions. Just as self-care increases one's benefits or health and well-being, considering the welfare of others, whether kin or members of a small group, including one's habitat, likewise increases advantages.

Cooperation is not a uniquely human behavior, as mutual aid (Kropotkin, 1902) is freely given and distributed among apes. Research in the field and lab demonstrates that great apes recognize fairness or sharing and punish cheating or subversion (e.g., see Sarah Brosnan, this bibliography). Why are apparently human-only attitudes like respect and loyalty also evident in apes? Simply, these behaviors, related to altruism and cooperation, what people might call *moral*, were important for genetic preservation and perhaps environmentally driven. Importantly, then, the moral individualism of the ape is closely tied to the ape ethic regarding the maintenance of the range in her habitat. Apes are intimately connected to and depend upon each other for survival and reproduction in an organic environment.

Roughly, let me apply ethologist Niko Tinbergen's (1963) criteria for the evolution of behavior. Note that there's also some reliance on biologists Verne Grant (1963), Konrad Lorenz (1965), and Jerram Brown (1975). Tinbergen would consider the following if positing an ape ethic of moral individualism as eco-engineering. How the standard is met is stated.

1. Function. From personal self-care and kin welfare arose concern for habitat and group;
2. Phylogeny. Selection favored care and maintenance traits that proved advantageous;
3. Mechanism. Cognition and its process outcome, intelligence, preferred sustainable attitudes toward others and habitats;
4. Development. Though the drive for prospering is innate, environmentally a moral individual and her ethic must be cultivated through a long nurturing process from mother and group members.

Any adaptation justifies itself with reference to competing adaptations. Is there any strong adaptation in an organism that seems optimally evolved? The answer is no. That's too conditional, predictive, and not empirical, says Andy Gardner (2017). Following Ronald Fisher, Gardner says that organisms maximize their fitness (for survival and reproduction) in a way that makes it appear there is some purpose involved. This perspective shows that ape behaviors like eco-engineering, including sentiments of approval/disapproval, are not designed but subject to natural selection in accordance with Tinbergen's principles. Drawing from Williams (1966), adaptation is a physical or behavioral trait or process advantageous to organisms as they compete for homeostasis with conspecifics in their shared evolutionary environment. This selection creates a feedback loop where the social and ecological behaviors reinforce the selection process itself. Hence, apes evolved a land ethic of ecosystem engineering in their environment of evolutionary adaptation because it proved beneficial.

All in all, following Tinbergen, the naturally evolved social and ecological behaviors of great apes seem to make better evolutionary sense (i.e., a land ethic) than the culturally evolved aggrandizing and destructive tendencies of most modern human societies.

Indeed, many people willingly exclude most of the natural world and fashion themselves "humanists," as if that's a badge of honor. Responsible action will not happen on a widespread scale, and thus the appeal for ape personhood. More specifically, there's a call to recognize great apes as persons of the forest, moral individuals who act responsibly as ecosystem engineers. Taking an ape out of the forest and confining or raising him in a cage (like Harambe) renders him, no puns intended, a sitting duck, a fish out of water. Authors like Richard Cupp (2016) assert that personhood arguments are dangerous. Why? Because, according to his line of thinking, a speciesism argument at base, granting personhood to other "animals" dilutes *humanity*. This, too, is an act of (human) self-deception or, as biologist Robert Trivers (2011) might say, the folly of fools. As a side note, according to seventeenth- and eighteenth-century thought, the *Scala Naturae* conveniently had species fixed on a natural scale, meaning that while there might be obvious similarity between apes and humans, the apes were permanently and conveniently fixed under humans on the chain.

One does not choose to live in a group where others will deceive her; deception is simply a consequence of the group, ranging from basic forms in great apes, to insidiously malicious layers among some humans. Because of this likely scenario of self/other deceit and detection the notion of moral individualism can be assigned to apes who certainly deceive each other, and even humans, to get what they want. Lori Gruen (2019) tells the story of one chimpanzee holding the mouth of another so as not to reveal to the group the appearance of prized treats. The signing orangutan Chantek would intentionally deceive with signs (Miles, 1993). For example, Chantek would sign that he needed to use the toilet, but he'd go there to play. Or he'd sign that he wanted to play with a stuffed monkey toy, only to get into that room since it had tools he wanted to experiment with. Chantek had many toys and would play with them as might a four-year-old human child. These are innocuous deceptions that point to Chantek's moral individualism. Don't forget that "moral" does not necessarily mean to act ethically at all times; it means that one understands the difference between approved and disapproved behavior.

Many great apes are like a young human child, sensitive and perceptive; but who would deny personhood to children? The metric is erroneously human. A mature ape is an adult, not a human child. Humans are genetically similar to great apes, mostly to chimpanzees and bonobos, but very close to gorillas and then orangutans. Yet clearly some human DNA, which turns certain genes on/off (i.e., in the regulation of higher cognitive functions) marks a difference. Humans, like great apes, act from biological imperatives (e.g., competition and cooperation), and human social groups are larger and more densely built. While vervet monkeys and baboons can vocalize about a specific type of predator coming from a certain direction, chimpanzees share some human body language, though our linguistic abilities are more sophisticated according to a human measurement. Nonetheless, there are many examples of how great apes are taught to use sign language or lexigrams, so they indeed learn and understand complex communication other than their own. The bottom line is that human beings have evolved from some common ancestor shared with the chimpanzee and bonobo, and then, going back further, from the gorilla and the orangutan, all of whom adapted moral individualism to accommodate social and environmental niches. Jeffrey Schwartz (1987) maintains, on fossil and not necessarily DNA evidence, that there is a much closer evolutionary relationship between orangutans and humans. Chapters 2 and 3 will delve into species comparisons and contrasts.

Nevertheless, human personhood, however defined, is derived from previous hominid forms. My point is that for modern humans, the land ethic does not seem as centrally important as it is for great apes. There are continuities across species, and the similarities in physiology, mentality, and sociality still exist between humans and living great apes. This is not to overlook differences, evident in the land ethic. There is also, one could say, some fundamental similarity in terms of moral psychology, though applied differently between humans and apes. As Aristotle indicates, if there is no relevant difference, individuals should be treated alike (Rachels, 1993). What I'm aligning between apes and humans are cognitive capacities, intellectual outcomes, and moral social behaviors; what I'm differentiating is how most modern humans sully the environment while apes work to keep it pristine and functioning well.

From an ethnographic analysis of sixty societies, Oliver Scott Curry (2018) says that, cross culturally, morality is based on cooperation and some general concepts that cover kin care and group solidarity, reciprocity and protection of others, status recognition, and possession acknowledgment. Although Curry is talking about human morality and perhaps overestimates cooperation, it will come as no surprise that these very principles are demonstrated in great apes by virtue of their intelligence and social behavior. I'd suggest, moreover, that for apes, some of these "values" apply to their environs since their fitness is, on some scale, shared with and measured by their habitats.

Many species do not need a legal precedent or statute from humans to help them distinguish their differences or to assert their rights to life and liberty. Personhood is not an epiphenomenon but resides in the human genome common to apes and other primates going back through time. Some philosophers are uncomfortable with biological evolution since they think it means our behaviors are determined and that we have no free will. Nothing could be further from the truth (Tague, 2016, chapter 2), since mechanisms of natural selection ensure that there is variation in diversity for survival and reproductive success. That means that human individuals, because of unique phenotypes, bear and often exercise the burden of moral responsibility. Why not call that responsibility personhood among great apes who often exhibit quite a bit of self-restraint, group control, boundary recognition, and avoidance or reconciliation when confronted with social conflicts. While great apes and many animals evade aggression and violence since there are physical risks involved, self-command is deliberative and not determined. On the other hand, many humans tend to be physically or emotionally aggressive and violent to others, including strangers, for no apparent reason and with little or no real provocation. If we are *Homo sapiens*, the supposedly rational species, why would we take more risks and often inflict more harm, especially to our ecosphere, than our nearest ape cousins?

Indeed, what makes humans different from a species genetically close, that is, whether the chimpanzee or the orangutan, so that people engage in legal disputes over the definition of personhood on their behalf? This controversy explains the need for humans to rewrite their story of evolution. While rudiments surely exist in great ape behavior, humans have sophisticated mental narrative. Human consciousness is such that people construct the story of their own lives and are eager to enter into the lives of others through story. Laurent Dubreuil and Sue Savage-Rumbaugh (2019) show how bonobos can do this, too. Philosopher Thomas Nagel (1979) has famously asked what it is like to be a bat. As Paula Droege (2012) is right to note, pertinent to my argument, it might be more appropriate to ask what it's like for an ape to perform a certain behavior in a specific tropical environment at a precise time. This latter question takes into account autonoetic consciousness, or the ability of the ape to use past experience in achieving a future ecological goal that consists of temporal representations from memory traces.

Frans de Waal (2016a) does not, contrary to Nagel, believe it is impossible for us to imagine the inner, micro-world of an animal. Other species, as Todd Feinberg and Jon Mallat (2016) explain, have basic conscious awareness, but do we know what it is like? Does it really matter that we know since each species has evolved its own way to fill a certain ecological niche. We can hazard an educated guess about animal mental states since conscious awareness expresses content, meaning, and the phenomenology of subjectivity or what it feels like. Primatologists Dorothy Cheney and Richard Seyfarth (1990) have gone as far as suggesting that baboons theorize about another's behavior. While the debate about ape theory of mind has seemingly been settled by Christopher Krupenye et al. (2016), human theory of mind appears more advanced and lends itself to personal narratives for social advantage. Apes, too, engage in such moral mentality, evidenced from the narratives constructed by signing apes.

Philosopher Jeff McMahan (2005) supports Rachels' notion of moral individuals across species without placing higher value on humans. Indeed, human cognitive and psychological capacities should preclude maltreatment of potentially equal beings. The question is whether moral individualism places value only on intrinsic qualities or includes external characteristics, such as the relation of one individual to another or of the individual to her living environment. McMahan sees moral individualism squarely centered in intrinsic qualities and characteristics alone, many elements of which are shared between humans and animals. Yet people ignore or discount the moral individuality of other creatures, evident most prominently but certainly not limited to, those confined in zoo enclosures. Should membership in the human species confer moral superiority over members of other species? I don't think so, nor would, it seems, thinkers like Tom Regan (2004) or Andrew Linzey (2009). The biology of being human is not of itself moral-granting any more than the biology of, for example, an orangutan feeding with her baby and other orangutan mothers in rainforest trees.

Objectors can't simply say, putting aside biological elements, that there's some special glue holding humans together, for there are many social and cultural aspects to great ape groups, concrete examples of which appear in this book through chapters 2 and especially 3. Species membership and shared interests are patently not reserved for humans alone. As Jane Desmond (2012) points out, for instance, intentional art making by apes contests humanism. Objections to moral individualism point to a supposedly extended "common humanity." Rather, the ape ethic of moral individualism propounded here empirically suggests their inclusion in and relation to the environment of their evolutionary adaptation, which includes other apes and specifically the ecosystem habitat on which they depend and for which they seem to care. Apes, particularly orangutans, appear to exhibit moral states concerning the status of their environment, as if both mind and nature are coterminous domains. Can I prove this beyond the shadow of a doubt? Do I have to? Given the current state of climate crisis, I raise some reasons why great apes should be left alone to continue caring for abundant expanses of tropical forests that replenish the earth's atmosphere with stabilizing effects for the oceans.

Nicolas Delon (2015) challenges moral individualism, mostly because it is simply taken for granted by animal rights supporters. Delon says that some intrinsic qualities are indeed necessary for a creature's fulfillment. However, his emphasis for considering moral status lies in relevant "extrinsic properties." Moral individualism calls on one to recognize the essential moral status of another. The stress in *An Ape Ethic*, though, relies on moral individualism as an active state, not a conferred status, and so includes external properties, like the ape land ethic of eco-engineering. In other words, the intrinsic properties of great apes matter because of their often mutualistic relations to the environment of evolutionary adaptation, which includes other apes as well as the ecosystem of plants and multiple organisms. In this way, the argument here for moral individualism reveals how extrinsic features and characteristics are certainly relevant to the moral state of mind of the individual among other, like-minded agents. The point of this book is that because great apes are moral individuals they confer value to their habitat. Their moral individualism is not just a static, human-granted status but evident in them as dynamic agents of change and concern in a wide, ecological context. Delon says there are agent obligations as part of status and not merely intrinsic qualities. In discussing moral individualism, due care will be given to examining the research on great ape intelligence (intrinsic qualities) and social behavior (extrinsic qualities) as part of their land ethic.

GREAT APES AS ECOSYSTEM ENGINEERS

Let me move beyond this debate and take a slightly different but related perspective on the issue of great ape moral personhood. Many animals and mainly great apes can exhibit rational behavior, where thinking produces reasoned decision making or problem solving in the ecological as well as the social sphere. As philosopher Kristin Andrews (2016) notes, such rationality knowingly results in causality. Actions have teleological probabilities. In addition to self and bodily awareness, one could posit, then, that great apes are likely ecologically conscious that their foraging movements have consequences. They pick fruit, eat, and upon return at a later time find less/more fruit there. Functional rationality holds mental representations that act as propositions, after all. The ape mind is capable of making decisions about range roving and resource allocations of full/empty and plentiful/scarce. Great apes seem to be deliberate cultivators. For example, beyond instinct and demonstrating an environmental intelligence, orangutans plan routes for food, have toxicity knowledge, show flexibility in locating and processing foods, and engage in nutritional variability (Galdikas & Shapiro, 1996). Darwin (1859, chapter 7) says that while "instinct impels" there is, surely, reasoning on a scale with animal ecological intelligence, as my book is trying to establish.

Forests, as that word is used loosely here, in Borneo, for example, help regulate water supplies to surrounding regions, and these massive areas of mother trees, their offspring, and other vegetation are critical carbon sinks. For apes, the forest integrity of the trees they rely on for feeding is vital to fighting climate changes, suggesting a coevolved mutualism of ecologically sustainable interdependent species (Lee et al., 2019). Since orangutans move freely about various forest types seasonally, they contribute to biodiversity. In *Abundant Earth* Eileen Crist (2019, p. 14) defines biodiversity as follows, and this is particularly apt for combined points I'm trying to tie together about ecosystem engineering. Biodiversity is more than mere numbers of species and includes "food webs" and "nutrient circulation." Any so-called over harvesting of the forest's resources stems from anthropogenic activity, not from the apes themselves now or in their deep history there. Since orangutans in unison with other species in their ecosystem are all purveyors of rainforest growth they, in consequence, contribute positively to the effects of subsequent rainfalls. In other words, orangutans seem to have evolved as enablers of overall tropical forest health represented in its trees.

The process of ecosystem engineering refers to any organism which not only creates but also modifies its habitat, what Crist (2019) might refer to as impregnating dynamic energy into the biosphere. Anne Russon et al. (2008) submit that orangutans do not deforest their habitats but aid large alluvial areas like swamps to promote fruit growth. According to Clive Jones et al. (1994), key to ecosystem engineering is how an organism modulates the supply of resources. Beavers are eco-engineers. It's well known that organic matter excreted from a hippopotamus, who spends much time in water, feeds nutrients to fish, insects, and hence land animals that eat those species. In other words, a habitat might not only be determined by food resources but by an organism's ability to maintain the health and stability of the locale and consequently affect the distribution or abundance of resources to other biodiverse species, like plants. While many species are involved in niche construction to a certain degree, the real question is which species exert a significant enough force on the physicality of the ecosystem to affect it dramatically. Apes and other species promote forest evolution whereas modern humans have seriously hampered such growth. Darwin was aware, for example, that earthworm activity on a fundamental level plays into soil formation. My claim is that great apes are super ecosystem engineers since they impact large forests with global climate consequences. Taking away the forests apes diligently attend to results in less beneficial effects on the climate. Certainly not alone, but great apes might be considered principal climate controllers, in a sense.

Ecosystem engineers alter biotic and abiotic matter, which in turn affects resources for other organisms. Eco-engineers have the ability to change surroundings and are called engineers because of their impact on an ecosystem as they modify a habitat. Joshua Barry et al. (2018), for instance, relate the case of pumas in the western United States who kill large prey, eat only so much, and then leave a carcass that becomes home for over 200 beetle species which, in turn, break down the biomass for smaller organisms, all trickling into soil composition. A similar pattern, with fruits and leaves, happens with orangutans in a rainforest. Clive Jones et al. (1997) say that ecosystem engineering can be direct or indirect control of resources for other organisms via "modification, maintenance, or creation of habitats." Eco-engineering does not necessarily have to include "feedback effects" to the engineer, and while there can be negatives in a small, confined area, ecosystem engineering would overall be positive in large areas like rainforests since, according to Jones' thinking, it controls the "flows of energy." Whether conscious or not, open to debate, this engineering cycle is surely an evolved adaptation for habitat sustenance.

The problem with dismissing ape ecosystem engineering as nonconscious is in how critics incorrectly measure systems. The consciousness in eco-engineers is less individual and more distributed, evident in how all parts of the modular distribution move to form a contiguous and continuous whole. A loose analogy might be in how either a flock of birds or a school of fish swarms for food or protection where control awareness is shared. In essence, without simplifying or eliminating individual agency, there is decentralized but naturally occurring coordination in ape decision making. In other words, collective behavior to produce a stable output, according to Bryan Daniels et al. (2016), includes amplification, or how different sized scales react to each other, and decomposability, or how individual components and multiple synergistic elements aggregate.

R. D. Haemig (2012) reports on two types of ecosystem engineers:

1. Allogenic, with change caused by transforming another organism's state;
2. Autogenic, with change caused by the physical engineer itself (e.g., its living or dead body as home for new organisms, like a downed tree).

Of course the question of intent arises here, for much eco-engineering is a byproduct of other conscious activity. However, the byproduct is part of an ape's total, physical awareness of her environment. Beavers, for instance, are examples of allogenic engineers, for in the action of their tree cutting and dam creation the local environment is considerably altered. The subsequent pond created by the dam becomes home for new amphibians, birds, and insects. Likewise, great apes are allogenic actors when they disperse organic matter, like leaves or discarded fruit rinds and cores, and especially seeds over a wide area. Crist (2019) flatly states that rainforests are in peril because of the loss of seed dispersers, like primates. Nancy Conklin-Brittain, Cheryl Knott, and Richard Wrangham (2001) say that for orangutans, granivory or seed predation should be a feeding category separate from frugivory. Crucially, these authors go on to say that spacing among orangutans while eating lessens the pressure on fruiting patches. Such behavior helps apes redistribute growth-potential material and create microhabitats for certain organisms encouraging new plant flourishing. In turn, discarded, partly eaten or unused foliage dropped to the forest bed becomes part of allogenic and autogenic engineering. Orangutan feeding habits can increase the biomass since fallen organic matter prompts other, smaller species to till the surrounding soil as they feed. Abandoned ape sleeping nests, whether on the ground or in trees, become miniature habitats for insects, which in turn attract other species who feed on them continuing a cycle of resource growth, decay, and energy distribution.

Orangutans in clearing dead clutter from above will then promote the fallen branches to become home for plants which grow on them, epiphytes. Liana wood vines are examples of autogenic engineering, since they alter the environment by modifying themselves and facilitate primate tree crossings. Haemig (2012) notes that ecosystem engineers, by virtue of their distribution of materials, could decrease some species, but in terms of the larger scale eco-engineers, like orangutans and other great apes, a "more heterogeneous" biome is created. Though difficult to measure, in theory ape ecosystem engineering, because of feeding, nesting, and branch clearing, could create local conditions not normally present and consequently establish a new home for species that would not typically present at that spot. Importantly, it's not as if apes "engineer" the entire forest. That's not the point. The engineering could even occur on a chemical level, where excreted bodily fluids promote microscopic organisms which, in turn, have a positive growth effect on the forest floor. The point is that the presence of and the adapted actions by apes instigate tropical developmental effects on several scales. There are many plant and animal species as well as microorganisms that need to construct their own habitats undisturbed. The crux here is that evolved ape activity increases biodiversity through a harvesting of resources without unduly suffocating any species in a locale.

Apes are big species or megafauna who range over wide areas and have an important impact not only to modulate the distribution of resources for other species but also to help generate new structures through seed dispersals. This is not haphazard but patterned behavior. In terms of natural selection, it is the evolved behavior of an ape that benefits her fitness or that of her genetic progeny. Any so-called engineering is of course an analogy, but the design aspect of the eco-engineering in an ecosystem clearly benefits the designer by eliciting interactions with other organisms in a bio-network and subsequently the promotion of food resources for survival. Is there, then, intent in ecological system engineering behavior? Justin Wright et al. (2006) say that when the term "ecosystem engineering" was first coined by Clive Jones et al. (1994), intent was not part of the definition. Rather, it is a useful process employed by some key species to modify and distribute potential and consequent energy growth advantageously. This is not to minimize some species whose eco-engineering is subtle but widespread, like plant-pollenating bees. The purpose here is to show how ape ecosystem engineering is part of a suite of behaviors, in line with cognition, intelligence, and social behaviors entitling them to forest personhood and dominion. On some primal level they sense (viscerally feel) what they are doing in order to survive. The engineering might not be overtly deliberate according to human standards, but it is conscious behavior, much as competitor avoidance or group formation. The problem with the question of intent is that humans look for defined outcomes where the apes intuit multistep net effects, as Wright might say. In spite of any debate over outcomes and intent, apes surely act like eco-engineers.

Darwin (1859, chapter 11) discusses the transmission of seeds, submerged or suspended in water, and how dried seeds can float for months and then germinate. Multiply that process by many species over long expanses of time. Even with such seed migration, as Darwin says, there are, nevertheless, distinct areas or ecosystems, as in a rainforest. Darwin, furthermore (1859, chapter 12), directly states that nature is a gardener. He implies that mutualism and competition among species help evolve and manage an ecological niche of resources. Indeed, Crist (2019, p. 49) says that cooperation in its broadest sense is not only human since behavioral diversity is a "strategy for avoiding competition." The more varied the habits and needs the better various species can survive together. At the same time, however, desert rattlesnakes, say Randall Reiserer et al. (2018), are secondary seed dispersers although carnivores, since they eat birds or rodents who carry seeds in mouth pouches. Moreover, while the integument of fruit is an adaptation to protect seeds during mastication, the flesh of fruit is also an adaptation to attract seed predators, like apes, who assist as primary seed dispersers.

Patterns of orangutan movement, affected by the physical contours of a forest environment, ultimately increase tree spacing via seed dispersal (Blackburn et al., 2019). For many modern humans so estranged from the activities of the natural world it's difficult to separate process and outcome. Hunter-gatherer societies would likely understand this mutual process. As Yuval Harari (2017) notes, hunter-gatherers in small groups did not negatively affect their ecosystems and were, therefore, unaware of environmental impacts. I'd like to draw a parallel to apes by saying that this question of awareness depends on scale. Like apes, hunter-gatherers need a large range so that small groups sustainably fit into an ecological matrix with other groups and other species. In the natural world the eco-engineer engages in a dynamic exchange that results in consequences for other species, which may ultimately return to the engineer as part of the process. I've already referred to this cycle as, by analogy, inclusive fitness. Essentially, in ecosystem engineering theory, there is an evolutionary operation of organic material redistribution in niche construction with positive feedbacks across species and especially to the engineer. Orangutans don't landscape for its own sake but cultivate a sustainable forest garden for their advantage and, as a result, for the benefit of others. This attitude is in contrast to the often curious human fascination with gardens as aesthetic objects.

For instance, biophilia is the human love for all life forms, at least by definition but surely not in practice. Social psychologist Anthony Rose (2011) says biophilia is merely a manifestation of the modern human need for hierarchy over other forms of life. I agree with Rose that somehow when we look to "nature" we reflect back human needs and desires. We recognize the value and beauty of animals, says Carlo Alvaro (2019), but we are conditioned not to rely on this feeling, making us morally blind to persecuting them. Most humans don't *really* love other life forms. Many societies separate themselves from other organisms and plunder natural resources for economic gain or political means. Modern humans have created pollution factories that raise animals for meat consumption. Rather, says Rose, there should be a full application of biosynergy, which is a communal exchange of resources among species in harmony. In other words, the biosynergy bond between humans and nature is preferred since it is not, like biophilia, detached or even competitive. Rose says Erich Fromm's original 1964 definition of biophilia included "love for humanity." The fact that love for humanity, an emotional appeal, is included implies that the focus is entirely on the needs of humans, a logos appeal. If one's focus is entirely on the human, then other life forms could suffer insofar as they can serve or satisfy people. Modern humans seem to compete against nature on behalf of humanity, while apes evidently cooperate with nature for the health of tropical forests, which, in turn, are immanently advantageous to global climate health.

Where biophilia excludes, biosynergy includes (Rose, 2011). Essentially, some of *An Ape Ethic* is on the order of biophilia, for it satisfies the human curiosity to study apes without, in itself, interacting with them reciprocally. By their nature apes are biosynergists. Nevertheless, why do many humans always look at these situations from their own perspective and not from that of the ape's? Stacey Sowards (2006) suggests how a human can identify with an orangutan so as to have, in perhaps Nietzschean fashion, a revaluation of *human* values, an interspecies event. Sowards labels this experience as consubstantiation, even if only rhetorically, to help people understand how to break boundaries and understand what it is like to be an ape. For all we know, this bodily intersubjectivity of *feeling what it is to be like* occurs for apes as well, evidenced in the story of Judy at the opening of this chapter and in conversations with apes by Laurent Dubreuil and Sue Savage-Rumbaugh (2019). In their effort to imitate everything human, orangutans very often take a human point of view. Human personification of apes is natural and even necessary in this regard of grasping their personhood.

In the final essay of his book, *A Sand County Almanac*, American conservationist Aldo Leopold (1949) calls for a land ethic stating that forests and open spaces should not be used for economic expediency that inevitably leads to degradation. Instead, use should be considered as a matter of environmental wholeness, an ethic of land integrity. This commendable stance is, nonetheless, closer to biophilia than biosynergy because it looks at nature through human eyes. How does nature look to an orangutan? Both species do not see the same resources available for similar fitness purposes. In his great literary epic, *Paradise Lost*, John Milton tells how the human-like fallen angels in hell dig into the earth for ribs of gold to build Pandemonium. There's the difference. Great apes have earned the right to exist in their natural habitats without being unduly disturbed. They are not fallen angels who act like some greedy humans despoiling the earth. Leopold notes how people readily accept a philosophical ethic that makes a distinction between what is good for society and what is not. Similarly, a land ethic calls for boundaries on how land is managed. Primates inhabit forests under attack from governments, local people, and multinational corporations, whether in tropical Africa, South America, Asia, or Indonesia. Whereas land, previously, was viewed as property without obligation, that can no longer be the case. Why? As Leopold notes, there are many plant and animal species in peril when the land is held only in terms of human economic value.

Leopold is correct to say most people are self-interested, for where would any human or any ape be without some egocentrism? But overwhelming self-interest can become destructive especially when held by powerful groups. This attitude is evidenced in continuous human population growth and insatiable resource consumption and product manufacturing in many, but of course not all, societies. Regarding land use, ape groups don't seem to share this often negative human group ethic. An ape ethic is not based on human economic value that voluntarily destroys habitats but on what is best for the different species to thrive in their biotic communities. Moreover, an ape ethic says that these creatures clearly exercise a land ethic, the sense of right over expediency, through their successful evolution in filling their distinct social and ecological niches. Apes are not destroying their own habitats but work in balance with the environment, evidenced, for example, in evolved controls on population expansion by long birth intervals and single births. Ranging, along with activity budgets and diet, across a large spaced feeding area is a key component to successful ecosystem engineering. To restrict frugivorous species like orangutans geographically to so-called parks or preserves means they will possibly be forced to deplete resources. As part of their land ethic, distinct but contiguous groups of apes evolved, as a whole, to roam fairly widely.

More precisely, great apes, like early hominins from *Australopithecus* to *Homo erectus*, exhibit a moral individualism in their land ethic. If we count these early humans as "persons," then logic dictates that we should do so for great apes as well. Through smaller populations they contribute to the balanced sustainability of their ecosystems. We know apes have an ecological sense, their own original biosynergy, because once deprived of it, if forced into captivity, they languish. Apes value freedom in their own green habitats. Those who fail to see great ape *mores* as the expression of moral individualism and not only an adaptation disregard human morality as evolved (Tague, 2016, chapter 2). Although it will make some readers uncomfortable, I question basic assumptions about the moral lives of great apes. If humans are so morally advanced, why are many usually in states of conflict and war against each other and the environment while apes typically, though not always (depending on the species), live in peace within their own communities? Given the opportunity, would great apes deplete their own habitats of food? Based on their evolutionary history they have not done so. *Homo sapiens* is a relatively young species yet has willfully corrupted not only many parts of the earth but also the homes of other species to the point of forced extinction. From my accounting so far, it appears that nonhuman primates exhibit a land ethic superior to that of humans.

Are great apes stewards of their environments? Sustainability is not just for those humans engaged in pillaging and polluting economies; great apes need a sustainable ecosystem in order to survive. When people look at sustainability, they tend to look at it from a human perspective with the question, *How can it benefit society?* Rather, people need to see how ecosystems must remain stabilized so as to permit the many species there a sustainable habitat. After all, humans are not the only species with societies. Chimpanzees forage, use tools, and gather materials to build nests, but on an ecological scale that does not negatively impact their territory dramatically. Cheryll Knott (1998) studied changes in orangutan ketones, which metabolize fat for energy. During periods of poor fruiting ketones were high, suggesting, perhaps, that orangutans efficiently store fat for later use, an evolved adaptation as part of eco-engineering in terms of regulating the body in parallel with the environment. On the other hand, humans raze woodlands to build "livestock" warehouses to store their "fats," and these factory farms create extensive environmental waste through loss of arable land, agricultural products used to feed cattle, offal runoff that contaminates water and soil, and the release of high levels of methane gases.

Based on their evolutionary trajectories and primitive technology, great apes are unlikely to destabilize their environments. Consider, again, the low birth rates and very long childhood development of, in particular, orangutans and mountain gorillas. Like hunter-gatherers, on some level they are aware of the contents and boundaries of their domains. Could their overall population equilibrium be an evolved adaptation that helps them maintain ecological homeostasis? Human overpopulation represents the major threat to earth's sustainability. Population density and body size in apes might correlate because of the environment of evolutionary adaptation to resources, evident in how a variety of species inhabit and survive in one locale. On the other hand, typically when humans develop an area for habitation, all the species are pushed out, and any remaining creatures are destroyed as "pests." Humans could learn quite a lot from ape intelligence and behavior. Mental maps help species find food and monitor competitive resource boundaries in an economical fashion without exploitation. Food resources impact the fitness of primates, like other species, to the extent that they have evolved into ecosystem engineers. Apes live healthily.

It seems fair to say that apes evolved to fill a certain niche, and because genes respond to environment, there has been a coevolution of morphology, intelligence, habitat, and culture. A principle of *Umwelt* or life-world (von Uexküll, 2010) emphasizes the shaping influence of an environment on both the conscious and unconscious minds, a cognitive aspect close to gestalt. While he'd not wholly accord with nongenetic stressors, Dawkins (1982), in discussing the extended phenotype and perhaps anticipating what's called the extended evolutionary synthesis, makes clear that genes even "manipulate" an environment to their own advantage. Take the case of self-medication. Is that ecosystem engineering? Some would say no, but engineering in its broadest sense means skilled. Here's a bit more detail.

Fiona Stewart (2011) raises questions about how nest building is an important ecosystem engineering component that led to the architecture and construction of home bases among hominids. This thinking is akin to Dawkins' (1982) extended phenotype where behaviors stretch beyond the body, impact the environment, and so affect the fitness of other organisms, all of which literally and figuratively feeds back to the apes. Consider, for instance, as have Adam van Casteren et al. (2012), the sophisticated biomechanics exhibited by orangutans in nest building where the activity is not random but tree-type specific. Barbara Harrisson (1962) offers a good description of how thoroughly a male orangutan builds his nest, with careful attention to detail and a combination of sequenced movements and thoughtful considerations. Similar knowledge of plant biology and mechanics are related by H.

C. Morrogh-Bernard et al. (2017) who studied fifty individual orangutans in the Bornean Sabangau peat swamp, central Kalimantan. They documented rare but somewhat continual use of *Dracaena cantleyi* leaves used to relieve muscle pain. Apparently, the local indigenous people do the same. This implies that on some cognitive level ecological niche construction and eco-engineering took place for use of the medicinal plant without diminishing it and encouraging its proliferation.

Surely this is a controversial idea. It is, however, well known and documented that primates are important to their ecosystems since they disperse seeds (Galdikas & Shapiro, 1996). Seed dispersal was studied in a quantitative, methodical manner by Esther Tarszisz et al. (2018) to imply a coevolution of orangutan physiology, movement, and ecology. Taking the idea of niche construction a step further, Karola Stotz (2017) suggests how organisms can engage in developmental niche construction. This means that beyond routine activity in a selective environment one is dynamic in and responsive to her developmental environment, which includes phenotypic variation and cultural pressure from parental, social, and even ecological influences. The constructed habitation of apes can be seen, in fact, as an extension of the phenotype. An ape extended phenotype is in the rainforest; the human extended phenotype, on the other hand, is primarily in built spaces and manifested in all of the stone, brick, concrete, steel, iron, and blacktop of that plastically constructed "environment." After all, Dawkins (1982) says a phenotype can instigate leverage across generations. Which of these extended phenotypes is more beneficial to carbon based life forms in the long run? Furthermore, in terms of ecosphere-level systems, Joris Cromsigt et al. (2018) acknowledge how megafauna plant eating species, like great apes, promote vegetation that withholds carbon while their feces and urine fertilize soil, stimulate further plant growth and, therefore, soak in more air carbon, ensuring the successful continuation of their home environments.

In their fruit and leaf eating habits, primates have selectively shaped plant traits and seeding, suggest Joanna Lambert et al. (1998). This is not far-fetched when one considers how much time in the beginning of *On the Origin of Species* Darwin (1859) talks about artificial selection. Perhaps the analogy of apes as artificially selecting plants is inaccurate, but it's a helpful visual. While primates are seed predators, this predation promotes seed dispersal in the handling, expectorating, defecation, and bodily movements across space. Is this deliberate ecosystem engineering? The cohabitation of plant/animal is not, says Lambert, coevolution since many seeds don't survive. True, but many seeds would not survive anyway, and certainly the advantage is more to the primates. While primates have not, according to Lambert, altered the evolution of plants, monkeys and apes are definitely an ecological force. Seed dispersal is massive, on the scale of hundreds of millions per year by many animals and insects. So primates and then apes have affected tree growth and distribution.

Does this mean that primates are ecosystem engineers? Colin Chapman et al. (2013) ask that question and answer affirmatively in how primates can modify their environments through this very seed dispersal. Primates can also eat so as to slow or advance the growth of certain plants, alter a species of plant, and increase the spread of soil nutrients. For example, plant tissue production and forest communities can be stimulated by primates who moderately graze and defoliate trees. In considering the impending extinction of some primate species, Alejandro Estrada et al. (2017) bluntly assert that primates matter to the sustainability and health of tropical biodiversity since they actively contribute to forest regeneration. How conscious is an orangutan of its forestry? In answering, let's say ape psychology is inseparably entwined in the natural world. Conservation is a complex and multistep approach to environmental problems, as Chapman et al. (2005) say, but clearly primates have been better weavers of the forest ecology than human destroyers. In light of environmental psychology (Barrett, 2011), to be discussed at the end of chapter 2, a breathing body evolved a brain, not a computer. A body with its regenerating cells is an ecosystem engineer since its *raison d'être* is for continuous nutritional sustenance and habitat sustainability.

Orangutans prefer fruit over leaves, but because there can be periods of low fruit resources, orangutans have to calculate how to survive. The strategy is to feed on an abundance of miscellaneous items. Without fruit orangutans travel less and to meet nutritional goals will eat more leaves, bark, and flowers reducing carbohydrate and caloric intake while increasing protein. There is, then, a metabolic adaptation for low energy. When fruit is available, though preferred, orangutans spread out protein intake by combining fruits and leaves. Plant leaves contain sugars, starches, proteins, and water. This buffet eating strategy assists not only the consumer but supports balanced growth across plant species. However, orangutans will overconsume fruits when available to store body fat during lean times. Regardless, if not consumed, the overly ripening fruit would rot and fall from the tree, piling seeds in choking competition at its own roots without dispersal. Metabolic adaptation might be considered part of ecosystem engineering where a primate's carbohydrate, protein, and caloric needs contribute to the relative prevalence of plant species. It has been shown that Arctic foxes, for example, eco-engineer their habitats for survival (Gharajehdaghipour, 2016), creating miniature food supply chains.

While this activity might not be conscious and merely a byproduct in the struggle for existence, the fact is that the adaptive behavior of stimulating plant growth by resource usage and seed dispersal has increased survivability over evolutionary time. Prime plant shoots inhibit the growth of other shoots, so it is common knowledge that pruning stimulates additional branch development. Shoot tips compete with each other, and there's a hormone called auxin in an actively growing plant that moves down the stem to inhibit growth of younger buds (Prusinkiewicz et al., 2009). Of course apes' eating reduces parts of plant species, but the point is that by doing so they force the plants to extend their growth in their own battle for survival and reproduction. Critics might say such action is not ecosystem engineering since it's not deliberate. It is certainly more subtle than when a beaver builds a dam, but the consideration of eating which fruits and leaves at which location is surely calculated. At the same time, note how conscious human activity has systematically, with purpose, reversed the health of ecological communities. Though difficult to measure, there is some level of agency in an ape's ability to thrive in a forest.

Regarding this idea of engineering, and loosely borrowing from Robert Heilbroner (1961) about economy that differs from sedentism and agriculture, great apes don't cut down trees but stimulate their growth in a codependency. In economic terms, roughly speaking, apes operate on a market structure where each individual operates for his or her own continued health in communal equilibrium. In spite of this freedom, apes have not pulled down the forests. Yet the human market system of capitalism wants to rule the forests. These two market devices are currently in opposition, with the one (of apes) demonstrating preservation and the other (of humans) representing destruction. This is not to say apes don't have cost/value concepts. While personal gain, a notion of property, and a sense of fairness are common among apes, they don't market in those tendencies as do humans. There's a supply balance.

Among apes there is no separate economic world from the ecological or social spheres. This is no small point. Great apes don't seek to advance their personal lot by accumulation but simply labor to maintain themselves through custom and tradition. If there is any nonhuman ape who systematically trades favors for gain and struggles for personal advantage in a community, it might be the chimpanzee. But even having said that seems to insult their ecological integrity. On average, ape economics is based not so much on transaction of personal property but on reciprocal interaction with the ecology and community through a primitive version of horticulture. Plants are magical in how they clean the air, attract life, absorb noise, and reduce stress. *Homo* evolved in a green environment but has since mostly lost communication with nature. Apes consciously continue to be forest persons, to wit, they have not left their forests.

At any rate, chimpanzees seem the most likely candidates to destabilize a locale. There is, for example, the instance of how chimpanzees in the Kibale National Park, Uganda, as reported by J. S. Lwanga et al. (2011), are excessively hunting red colobus monkeys. This monkey population has decreased over thirty years while the chimpanzee population has increased. However, to what extent might human actions have affected this outcome? After all, the chimpanzees and monkeys studied, along with all other animals, live in the restrictions of a national park that is less than 300 square miles, roughly the size of New York City. Human hunting of big cats or degrading their habitats has nearly eliminated them in some areas, and such feline predators among others, like bush pigs, could put a check on chimpanzee population. Besides, should anyone be surprised at this fact given how *like us* chimpanzees are? Since some great apes, such as chimpanzees, don't always live in utopian peace only shows how close they are to us. Their infighting often has less to do with obtaining food for survival but more to dominance and status in terms of who gets what and who shares with whom.

Equilibrium requires a large, spatial scale, which is what apes once had for the free movement of multiple groups. Human activities, however, can elicit detrimental selection pressures (Bekoff, 2013). People tend to examine everything from a human perspective in a world dominated by humans. Nonetheless, chimpanzees are closer to small hunter-gatherer groups who tend not to degrade an ecological niche. Of course I can only speculate from the evidence I've gathered about any possible ecological maintenance and, given the chance with more sophisticated technology, perhaps some apes would debilitate their habitat like humans. However, on some conscious, semiconscious, or adaptive level great apes express moral individualism in their land ethic, pointing directly to what humans would label personhood.

Humans pride themselves on their rationality. Like Leopold (1949) let's now call for a conscience of responsible conviction for the welfare of great apes by virtue of their conservation of lands for self-renewal. In that respect they stand apart from modern humans. It seems reasonable, based on ape-human continuities and similarities, to accept great apes as our evolutionary peers. Given the chance, would apes push humans off their land? That hasn't happened yet. In this way, great apes should be granted sovereignty over their forest habitats in line with thinking by Donaldson and Kymlicka (2011) and Meijer (2019).

INDIVIDUAL AGENCY AND AUTONOMY

A note of caution comes from the esteemed Robert Hinde (1987), the zoologist who trained Jane Goodall and Dian Fossey, in his assertion that there are dangers in making parallels between humans and primates, mostly because of the several hundred primate species and even greater number of human societies. Hinde asks that one not make inferences about the human condition from apes, but that one should apply a principle. First, there are only four extant great ape types. Second, three related principles will be applied throughout this book, as outlined so far:

1. Moral individualism
2. An ape land ethic
3. Ecosystem engineering

From this list number 1 is shared by humans and apes; numbers 2 and 3 are heightened in nonhuman apes. The question is which behaviors make biological sense. Suffice it to say, outside of similarities of morphology and physiology, humans and great apes share many behaviors across their social and psychological spectra. Yet apes, but not all human societies, appear to express a land ethic as moral individuals engineering their ecosystems through sustainable resource management. As one instance, much human activity is systematically killing essential pollinating insects on a wide scale.

In a discussion of autonomy, philosopher Wayne Williams (2015) says how a subject with values not only deserves respect but holds agency regarding those standards. An orangutan does not simply value the tree canopy but demonstrates a responsible attitude toward using it as a renewable resource. I don't need to prove this any further. In other words, great apes are not passive recipients of moral attention by humans but willful generators of morally responsible actions. Autonomy is best noted by Biruté Galdikas and Gary Shapiro (1996) documenting the semi-wild orangutans who enter and leave Camp Leakey as they wish, making deliberate decisions about whether or not to relate to and socially interact with humans and which ones in particular. Care must be placed, however, on how "autonomy" is used here. For example, philosopher Kristin Andrews et al. (2019) note that people often do not reason through moral decisions and can have automatic or emotionally based reactions. Therefore, autonomy alone need not be a determining factor of personhood but ranks high since it involves intentions and choices about what to eat, where to travel, and with whom to associate.

In applying a principle, philosopher James Rachels (1990) asserts that human beings are not in a special moral category. Furthermore, dignity is a concept manufactured by humans. What entitles humans to dignity? Are there not behaviors in great apes, like stature, athleticism, leadership, and intelligence that can classify them as noble? Humans are not divine or uniquely rational. Instead, what people think of as human dignity is rooted in the natural realm. Rather than some divinely rational dignity, then, Rachels offers the notion of moral individualism. There is no special class for being human, and how any organism is evaluated depends on its evolved traits and characteristics that fit an ecological niche. In this line of thinking, most humans must elevate near to their "level" the status of all creatures. Or many people need to come to the realization that all creatures are on an equal footing, in many respects, and entitled to the right to life in protected habitats. Among great apes and humans, it's less a hierarchical pyramid and more like a bumpy plane of shared equivalencies with some differences.

Abducting and imprisoning apes is, morally, unequal treatment, as is appropriation or misuse of their lands. Consider the brash irony in how Frederick King, director of the Yerkes primate center, published an article in *Science* in 1988 (Wise, 2000) that justified the capture and confinement of apes for experimentation since, as King plainly asserts, they are so much like humans. Nothing warrants the testing of lethal viruses on humans, so why on beings admittedly similar to humans? In fact, that's an admission of cruelty and an affirmation of ape personhood. They are like humans, says a key player, which paradoxically justifies treating them "inhumanely." Apes are autonomous, and both the repudiation of their liberty and the infliction of persecution deny their moral individualism. One thing is for sure. Semi-wild orangutans willfully and freely came to Gary Shapiro (1982, 1985, 1999) for company and communication, as he taught them sign language. Such unencumbered choice with clear intention implies they are moral agents. Very much like humans, apes prefer places, foods, objects, and others with an avid capacity to fulfill those personal preferences in their own habitats. At the same time apes are competent adjusters to any alteration of their natural environment in terms of needs and desires. They act in ways that demonstrate, beyond instinctual response or stimulus enhancement, abilities to understand and analyze their own and even another's choice.

Beyond the crass notion of animal kingdom there are individual agents like orangutans, gorillas, chimpanzees, and bonobos who diurnally act with purpose and, on reflection, revise such action for themselves, their kin, or their group. Could one go as far as saying great apes are phenomenologists? Richard Schmitt (1967) says that phenomenology is an exercise of reflection, but not necessarily about facts. Rather, the reflection is more about examples or how something is considered. To reflect is to bracket existence. To reflect is to question, and really to investigate one's belief of something. To be able to query implies a high level of cognitive awareness about the ability one has to question. Evident in chapters 2 and 3 that follow, apes are cognitively aware and abundantly responsible in a high degree, implying their ability to reflect and enquire about their own and another's belief. They can stand apart and bracket their existence from another's as witnessed by many field researchers asserting how apes recognize individual identity and personality. Each species is naturally selected and modified for different ends of survival and reproduction, so it is morally irrational for most humans to denigrate ape mentality and social behavior according to a human scale. That's not bracketing, for as Rachels (1990) points out, from Augustine, Aquinas, and Kant many people still cling to their supposed moral and rational superiority over animals. This is to say nothing about Descartes who, on grounds of human rationality, denied animal sapience and even sentience to justify vivisection. When I say sapience or sentience I'm not simply referring to deontology (right/wrong) or utilitarian (consequentialist) ethical theories but more to moral individualism (autonomy).

Philosopher Tom Regan (2004) questions whether "any animal" is autonomous as Kant has explained it: impartial, reason-governed action presupposing others in like circumstances would similarly act. However, because of their mental abilities in social situations great apes consider what others can or will do, and that is a moral stance. Additionally, moral philosopher Tom Beauchamp and colleague Victoria Wobber (2014) say there indeed is autonomous action in great apes (outside the confines of Kantian "rationality") absent a controlling force and evidenced in their self-control and intention using their knowledge in the context of an event. Another brand of autonomy, says Regan, emphasizes how an animal autonomously acts on preferences regardless of impartiality. In other words, there is a moral component to Kantian autonomy. Apes exercise self-control (e.g., chimpanzees in alpha trials) or restraint (e.g., male orangutans who recognize another's territory and keep distance), which are *mores* based on behavioral adaptations that nevertheless rely on individual temperament. One might say these are examples of self-interest. True, but then there are many instances of ape altruism which cannot be discounted and which clearly demonstrate how an ape acts for another *as if* he or she were in need. Outside of instinctual mother-infant bonds, reciprocal or indirect altruism requires a high degree of mental consideration.

In his chapters on the signing orangutan Chantek and the signing gorilla Koko, Steven Wise (2002) builds up to theory of mind and then ends with what he terms practical autonomy, or an individual's ability to make choices. In *Can Animals Be Persons?* Mark Rowlands (2019) also builds up to theory of mind, and while I place emphasis on mind reading in *An Ape Ethic* it's not actually the sine qua non of moral individualism. Apes possess intention and act with multilevel awareness. Leif Cocks (2016) tells the sad story of his meeting Chantek, who was then languishing in a research facility in Georgia. Chantek signed to Cocks to get the car keys so he could go home. This profound mentality demonstrates a high level of conscious awareness of his happy past, his present plight, and his imagined future. More so, Chantek demonstrated that he was keenly aware of his rights in asking for freedom. Autonomy evinces consciousness and deals with having a desire, intentionally acting on that desire, and having a high degree of self-awareness in knowing the desire and action are one's own (Wise, 2002, p. 32). Like human children, developmentally great apes mature through a combination of genes and environment, where environment is parental care and social interactions within a defined ecological habitat.

Consciousness and a sense of self are intimately tied to emotions, many of which we share with mammals. In consciousness one understands that internal states are connected to the world, and this capacity is evident in great apes who are able to sign emotions and who can, therefore, distinguish self from another through social and ecological actions. So in answering Hinde, the principle of moral individualism applies to apes from which their land ethic derives.

THE MORAL ETHOS AND AN APE ETHIC

The question of an ape "mind" was raised by Andrew Whiten (1996) and is linked to conscious self-awareness in relation to others as individual beings and as other minds. Self, however, is multidimensional (Hart & Karmel, 1996), consisting of memories, representations, and theories about oneself. Importantly, and a keen difference, perhaps, between apes and humans, is that for great apes the literal and figurative self is body in tropical environs. Self-reference and indexing among others has been documented, for example, in the orangutan Chantek, the gorilla Koko, the chimpanzee Washoe, and the bonobo Kanzi, to name only a few well-known apes. Of particular note, say Hart and Karmel, is how Michael, Koko's male companion, related through sign language the memory of his capture in Africa.

Natural selection has favored humans no more than other species which have, over great time, survived in their habitats. Yet people shove them out to certain doom. How can humans be morally special if so many inflict much needless harm and catastrophe on other species and their homes, particularly those who are closely related? Animals have proved their right to life again and again. See how they fight to stay alive. Are, then, animals sensitive to their intrinsic (genetic) worth? They have survived and reproduced in the struggle for existence over the billions of others who have become extinct. Nay—even those who have gone extinct have intrinsic worth since they provided some of the common, related ancestry of all extant living forms, plant and animal.

Like great apes who inhabit their own equatorial or Asian forests without harm to humans, people are capable of a higher moral ethos and so should let them remain as they are. What some humans perceive as moral superiority is, on an evolutionary plane, just another flavor of the same sort we see in animals. All species are different, yet many social species share some group norms. At the same time, there is shared ancestry in these similarities, not only differences, and that's the point Darwin made in 1859. Just as animals have social instincts, so do people, though humans eschew anything labeled instinct and call it their moral sense. Sensations and sentiments are closely related to what one can only call instincts. For example, we know animals exhibit altruistic behavior and will actually sacrifice to benefit another (e.g., Horner et al., 2011). As Peter Singer (1975, chapter 5) seems to ask, what is the boundary line for moral sensibility? Humans express moral feelings for kin and allies foremost, and then as the circle widens, with "animals" on the perimeter, humans lose their moral sense. Why should our moral behavior be limited to humans and not extended to other living creatures? Pets don't count since many pet owners eat animals and even abandon or harm their pets. There's no sensible reason to deny animals a moral sense, especially when one considers their evolutionary connectedness to us, their balanced ecological participation in the local and global arenas, and their social sentience and sapience.

So, as Rachels (1990) argues, moral individualism does not look to a special, dignified category for humankind but sees individuals across the natural world acting on their own account with similarities to other species. Humans should not silo off ethics and ignore the rights or deny the "moral" abilities of other species. Any just horizon ought to include the perception of animals as near equals. The line of thinking in my book, following Rachels, is crucial. We are not out to save the planet's environment only for humans. We should recognize that many humans have damaged the environments of other species, harming their moral individualism while they have not injured us. The remedy is to begin with an ape ethic, that is, recognizing their moral sensibilities to each other in their habitats, followed quickly by forest personhood and habitat sovereignty. Working for the great apes and their intricate ecosystems means we are working for human primates by default. Philosopher and biologist Bernard Rollin (1993) rightly speaks of respecting the telos of great apes; let me emphasize how their ethos is a model of moral individualism as ecological engineers.

Following philosopher Lawrence Hinman (2003), morality deals with foregrounded social beliefs and practices of the first order that actually guide behavior. Ethics is a second order in how one consciously reflects on moral conduct from the background so as to improve values and actions. Or, as philosopher Robert Holmes (1993) says, ethics is the study of morality. But like Hinman, Holmes does not see "animals" capable of assessing the worth of something or, therefore, ascribing goodness and establishing values. That's speciesism of the highest order. *An Ape Ethic* attempts to show, as I hope I've been doing so far, that great apes reason about their actions and their intimately connected ecological and social worlds since they can think abstractly. (More detail about their cognitive abilities comes in chapter 2 and moral behaviors in chapter 3.) They can momentarily set aside their own minds so as to imagine another person or place. Then they are capable of moral reasoning, that is, of considering alternatives of action. Yet the great ape alternative has not been to overtake communities but to seek flourishing in their home ranges unencumbered by human encroachment.

One can see that as highly social forest persons apes are moral actors, and because of their cognitive functions and intelligence there's no reason to doubt that they deliberately consider their own and another's actions for future adjustment. In the opening pages of his book, Hinman (2003) draws the comparison between nutrition and morality. There is moral health in how one tries to improve behavior as one would nourish physical health. The metaphor of mental nutrition for good behavior is especially apt for great apes who conscientiously do little harm in the expression of their interests and duties toward ecological and social habitats. Hinman is perhaps an old-fashioned philosopher and does not quite see that his metaphors of nutrition and moral behavior are evolved adaptations. On the one hand, then, offered here is an ethical theory about the way in which great apes structure their lives. On the other hand, this approach is closer to moral realism with reliance on the naturally evolved environmental behavior of apes in the appeal for ape personhood. It's difficult to consider the moral lives of great apes with any ranking in the fallacious great chain of being.

In the decades since the Green Revolution of the 1960s–1970s, much has been made from Aldo Leopold's (1949) environmentalism. Generally speaking, on one side of the debate environmentalists would sacrifice animals over land, while animal rights advocates would sacrifice land for animals. It's not quite that simple, of course. For instance, both sides agree that humans are part of the animal world and, therefore, not superior to any other species. The fact is that neither side can ignore the other's claims. As Peter Singer points out in *Animal Liberation* (1975), animals *should* be given moral consideration. Singer goes on to examine what is paramount in the entire animal kingdom: a being's self-interest and avoidance of suffering for itself and often for others. There could be some "moral" norms in those behaviors. Since animals are capable of enjoyment and avoid pain, one can say they have human-like interests. Indeed, humans have animal-like interests with appetites for food and sex. Singer says, then, that the boundary to mark others for moral consideration is not intelligence but the capacity for suffering.

Animals suffer, as seen in some of their behaviors, just as one can see how humans suffer. The mammalian nervous system is very much like a human's. Suffering from pain is an evolutionary adaptation, for once hurt an animal will hide as it recuperates. Beyond acknowledging moral status (Gruen, 2013), the larger argument is for the moral agency of apes, evident from their range of social behaviors. Moral agency comes in degrees and does not by its nature presuppose the highest attainment of ethical principles, says Paul Shapiro (2006). Denying moral behavior in animals, especially apes, is used to justify experimentation on them, slaughtering them for sport or consumption, and forcing their extinction. In talking about moral consideration, however, most humans still place themselves over animals. On the other hand, I do not see a faceless species but recognize individuals among the great apes who act with apparent reason and definite feeling.

This last statement, in part, flies against the thinking of philosophers like Kant and, it seems, Christine Korsgaard (Gruen, 2013) who place reflective reason only in humans. There is a section upcoming devoted to Korsgaard. Angela Coventry and Avram Hiller (2015) point out how Korsgaard, in following Kant and a strict reading of a small part of David Hume, seems to deny moral status to animals. As I demonstrate here and through an analysis of primate research papers in the next two chapters, great apes think and reason. Specifically, their mental activity is geared to ecology and renders apes as persons of forests. In her discussion of the moral status of animals, and taking into account thinkers like Peter Singer, Tom Regan, including Kant and even Korsgaard, Lori Gruen (2013) uses this apt expression: a "morally considerable being." This phrase rightly sees the moral behavior of great apes not as limited or negligible but as influential and important. Just as one culture can learn from another, for example, apply some hunter-gatherer principles to a capitalist economy, so humans can learn from the ape ethic of moral individualism in eco-engineering.

As philosopher Paola Cavalieri (2016) notes, it does not seem logical for people to extend moral protection to humans who are not moral agents, like children or those mentally impaired, and deny that protection to nonhuman beings like great apes who express agency, a salient point. The position here is not to stand aloft and discuss the welfare of animals or foist an ethical responsibility onto others. Rather, let's assume that great apes are moral individuals themselves, evident from their cooperative (even competitive), helping, caring, and thoughtful behaviors, as outlined in this book. Instead of insisting on moral status, which seems apparent given the evidence presented by others, let's embrace the moral states of apes from an empirical and not a metaphysical approach. There is no hierarchy of essences that separates humans from primates, as argued for by David Smith and Ioana Panaitiu (2015) in their discussion of dehumanization. There are, rather, species differences. Paola Cavalieri (2015) is right to criticize abstract thinking that minimizes apes for lacking a human essence and having only a subhuman essence. The old school sees moral status conferred on those who can morally reflect on their own actions, but this excludes those with serious cognitive impairments because of psychological disorders or advanced age.

Great apes ostensibly act as moral individuals and ecological stewards of their own welfare, but much human activity has jeopardized their ability to be just persons exercising an ape ethic. There seems to be little question that since most animals experience physical as well as mental suffering, they have a nascent moral sense. Why would someone believe that an "animal" who dies in the wild does not experience suffering? Mark Sagoff (2013) says so. There are well-documented cases, by Jane Goodall (1971, 1986), for example, of wild chimpanzees suffering depression and even death at the loss of an infant or mother. As if Goodall's word is not proof enough, some researchers (beginning with Harry Harlow in the 1950s) have subjected lab primates to maternal care deprivation just to make sure. And you can watch a video of me telling the sad story of the demise of Jimmy, a juvenile orangutan, from depression after separation from his adoptive mother (Tague, 2019). Personhood implies an understanding of rights and responsibilities, clearly exhibited by maternal care given and received. Some might argue this is merely instinctual. Then others could rebut and say that "instinct" manifests in human mothers too, not love. In fact, just as with humans, some ape mothers are not completely fit as maternal caregivers (Goodall, 1971, 1986).

Let's not simply ask for human moral obligations toward animals, but this is not to eradicate human duties to the natural world. Instead, let us maintain that some animals, for instance great apes, exhibit their own moral obligations to each other as forest persons. Humans are neither caretakers nor stewards of nature. On the contrary, modern humans usually degrade nature whereas other species, from algae to apes, maintain natural processes. There need not be a strict divide between environmentalists and animal rights activists. Why? In the animal world as creatures go about their business the debate between environmentalism and rights evaporates, absent the imprint of the human hand. Against the environmentalist's whole-scale view, individuals do matter. Think how infrequently orangutans produce offspring and the importance of all genetic flow among orangutans in any given areas, to say nothing of the inherent value of each individual.

Following Leopold, Sagoff (2013) says we have no moral duty to protect *individual* animals. Yet, consider the cases where cows who have escaped the slaughterhouse have been spared and sent to a farm sanctuary. Individuals do matter, as should be evident from my argument so far and especially with material I'll present in chapters 2 and 3. If many humans can lord over animals with the threat of death or extinction, how do people expect them to react? When humans say they treat animals humanely, that only means that they exert undue influence or control over them or slaughter them expeditiously. The right approach would be to have huge areas of land (well in excess of 300 square miles) off limits to humans permitting animals in these magnificent domains to live as they have for millennia. The planet is large enough to accommodate this environmentally decisive step. Eminent biologist E. O. Wilson via his Biodiversity Foundation suggests that half the earth be left alone in this way. Some political philosophers (Goodin et al., 1997) declare that great apes as self-determining beings with moral and political sovereignty over their habitats are entitled to possess their own forest nations. This line of thinking about animal dominion, which I clearly advocate, receives full consideration in Donaldson and Kymlicka (2011) and Meijer (2019). Furthermore, Herman Rijksen and Erik Meijaard (1999) say that because orangutans settled in Indonesia from Asia about 1.5mya (and might I suggest by extension all great ape species in other areas), preceding *Homo erectus* by about 1 million years, they have a "birthright" to the lands they've "managed" in their own type of ecological civilization. If an ape dies of natural causes in the forest, he has done so as a free agent with his own will and mind. Any great ape languishing in a zoo or restricted to a small scale government preserve is not the same as a wholly free-ranging person.

In a like-minded way, Dale Jamieson (2013) sees not tension, just differences, between environmentalists and animal rights advocates. The land ethic of Leopold and animal liberation are not, in the words of J. Baird Callicott "distinct and inconsistent perspectives" (quoted in Jamieson). Human values and practices are socially constructed. While humans have no doubt passed the point where, as just suggested, animals should be left alone, we can at least help people understand that animal rights are human rights. This last sentence requires a book itself, but put succinctly humans and animals do not live in separate spheres but on the same planet and are dependent upon each other. Isn't it alarming that there is now talk that suggests humans inhabit another planet, like Mars, given the possible doomsday scenario faced on earth. Why not take care of home? One small example is how with the loss of bees, through global warming and pesticide use, the human food chain is in jeopardy. As Marc Bekoff (2018) says, there are three interactive and connected pillars to global health: humans, other species, and the natural environment. Fundamentally, there is only one welfare, not many. What humans value should not be independent of how they experience and contemplate the natural world in which they live. Do humans think other creatures believe differently? One doesn't need a scientific paper to see that orangutans or mountain gorillas value their diminishing habitats. Just observe them in action. Humans don't really live in towns or cities but in an ecosphere that is connected to and dependent on regions far away, such as the deep oceans and tropical forests.

Since I've been discussing moral individualism, agency, and the ape ethic, I'd like to devote the next section to some neo-Kantian thinking from an important philosopher.

SPECIAL OBJECTIONS: THE LONG SHADOW OF KANT

This section deals exclusively with Christine Korsgaard's (2018) book, *Fellow Creatures*, and will attempt to do it justice. Hers is a robust and valuable philosophical treatise which I admire. It is dense and difficult, advocating for animals in a neo-Kantian argument but clearly favoring some human traits and characteristics. Philosophers like using Kant (rationality) in moral personhood arguments, but why not Hume (sympathy) or Schopenhauer (compassion)? Spotlighting Kant shrinks rather than expands the discussion. Far too much emphasis (even by me at times) is placed on the question of rationality. Instead, apes need to be considered as morally individual forest persons by virtue of the *Umwelt* of eco-engineering, the thrust of my book.

At the outset, Korsgaard inaccurately says animals live according to their nature, not values, because her argument uses "animals" generically and does not examine primates or great apes in any detail whatsoever. However, she admits that it is difficult to make human/animal comparisons since both have values of importance for themselves. Indeed, but then the remainder of her book does in fact make that comparison where animals, though admittedly not less important, are seen as having different needs justifying different treatment of them by humans. How could she say, if I'm not misreading her, that "continuation of life" *matters more* (7) to humans than animals because of vastly different degrees of relationships among members? Such a statement ignores the intricate hierarchical, coalitional, personal, and reciprocal long-term relationships of primates in a group and community. This matters because, by their nature, ape relations to each other include an ecological engagement with the forest. More so, philosopher Lori Gruen (2019) tells the amazing story about chimpanzee Emma, part of a group she rescued from a lab and relocated to a sanctuary. This young ape developed a relationship with Gruen to such a degree that at one time much later Emma pulled together and offered Gruen a gift of posy plants. Primatologists like Jan van Hooff also have stories about their extended relationships with apes who, after a lengthy separation, remember them and welcome them warmly (de Waal, 2019). Korsgaard seems to grant more but not absolute functional coherence of creating a self to humans rather than to animals. To some degree echoing Kant, she says animal "self" is more to instinct whereas human self is more to thought. I refer readers to my "Biology of Moral Systems" section in chapter 3 where I discuss the multiple selves outlined by Robert Mitchell (1994), all of which apply to great apes in some way. While Korsgaard, to her credit, is admittedly leaning to the case for animal rights and moral standing, it's hard not to see how often she privileges the human over the encompassing designation "animal" without pinpointing ape/human continuities.

In terms of rationality, Korsgaard wonders if an animal can comprehend the origin of beliefs so as to evaluate causes. Granted, that's difficult to answer, but will be covered here in chapter 2 in the discussion of Kristin Andrews (2015). Korsgaard seems progressive regarding animal prudence and autonomy, but at the same time moves ever closer to Kant to make her argument and not to primate or ape science. Primatologists from Köhler to de

Waal know, for instance, that apes can ponder their motives. Admitting this, though, implies that a strict Kantian must grant an ape autonomous personhood. Korsgaard reminds us that autonomy for Kant means the person regulates her behavior by moral principles she deems law as part of constructing a normative public self. This principle will be demonstrated by me in the many social subsections of chapter 3 and elsewhere.

However, Korsgaard believes this creation of self through decisions is only human. She does not think the morally evaluating self is sufficient, as per Kant. Rather, one should strive for a socially acceptable persona. Following G. E. M. Anscombe (1958) and with emphasis on virtue ethics, great apes are more than capable of social self-construction since they self-evaluate desires, motives, needs, and actions in a charged environment of conspecifics. Instances of this, too, will be outlined in the social sections of chapter 3. The argument against my line of thinking is that (supposedly) apes are not cooperative. In reply, show us a typical human (or human group) who is not often self-interested. Apes have a social ethic evidenced in their complex relationships, to say nothing of how they "cooperate" with their ecosystems. Examples of this, too, will be offered in chapter 3 and elsewhere.

Furthermore, in terms of harvesting and sharing food resources, one can see a land ethic of eco-engineering with deliberate if unconscious principles of action over instincts. That's not possible, some might object. In response, what about human dairy farming of the Neolithic era that produced lactose tolerance? Was that a conscious act to foster a flourishing milk industry? Human-only intention is not always grounded in reality.

Korsgaard might suggest that apes (again, she does not specify great apes but continuously refers to "animals") are imprisoned in a nonmoral world of instinct only. The human notion of moral aggression, likewise, can be seen operating in apes who gang up against one who breaks a socially held norm, such as stealing food or needlessly harming another outside of a dominance contest. This means apes, and not just humans, evaluate actions. In terms of the "species being" concept from Karl Marx, suggesting humanity's shared history, why does Korsgaard see that as a bonus? In spite of, or perhaps because of, how some humans bond there is rampant and endless damage to and destruction of other people, their homes, and the environment.

Korsgaard goes on to say that animals, unlike humans, cannot critically separate themselves from their surrounding environs. This notion of being fused with the environment is presented as something bad when it is good and bolsters my argument about the whole-body gestalt of apes in their biologically functioning *Umwelt* of interconnecting elements. Many human societies, on the other hand, in their separation from the natural world are destructive by fabricating pollution causing mechanisms and promoting ecologically damaging attitudes. Korsgaard mistakenly says the environment does not function for animals, but this thought disregards the cooperatively mutualistic evolutionary forces of organism and ecosystem examined in *An Ape Ethic*. See, for example, the difference between an ape's adaptive habitat and a humanly contrived garden (to be discussed in the Conclusion). Most human separation from the natural world is deliberate and detrimental, not a virtue as Korsgaard seems to suggest. Why is it a negative for an ape not to know of a different world, as she implies. Their world is sustainable enough for and by them.

The point of *An Ape Ethic* is exactly that an advanced organism like an orangutan is keenly aware of and synergistically responsible for her extended environment. Korsgaard claims that the natural world exists separately from human interests, but historically that is not true. Only now it seems so, an attitude leading to possible global doom. Animals have and express a vested interest in their homes, whether an earthworm, a beaver, an Arctic fox, or an orangutan. Nevertheless, to her credit Korsgaard has an entire chapter dealing with the case against human superiority, mostly because she sees how animals are important not so much in the grand scheme of Gaia but to other animals. She continues and, nonetheless, indicates how humans are superior through a supposedly advanced degree of morality, pursuant again to Kant.

In light of all the primatology research, why is there an ultimate appeal to a philosopher who never travelled to a jungle, who died barely into the nineteenth century, and hence had no exposure to the ideas of Charles Darwin? As shown in *An Ape Ethic*, apes are aware of the "moral" capacity for acts of approval/disapproval by self or others, for example, sharing or stealing food, aggressive behavior, or jealousy. Morality as social glue need not be conditioned on rationality; if one insists, then based on some of the evidence adduced in this book apes are indeed rational. Take the case of infanticide, which can occur among mountain gorillas. First, that genetic cleansing behavior is morally wrong by human standards. Second, it is, however, present among some human individuals and industrialized societies, though subtle. Ethics and morals are by their nature loaded with contradictions and not totally dependent on linear, static rational principles. At any rate, key here, in spite of these objections, is how Korsgaard in what appears as an astounding pirouette disagrees with Kant's belief that only humans have moral standing, though he was vehemently opposed to animal cruelty.

Some argue, says Korsgaard, that humans have no duty to animals because there is no reciprocal relationship. On the contrary, apes as eco-engineers of forests provide a service to the entire earthly biosphere to which all breathing creatures owe them a debt of justice. Apes much more than most humans are part of that mysterious self-regulating atmospheric and oceanic system Lovelock (1979) calls Gaia. Apes hold, therefore, a natural contract of sorts with their ecosystem habitats that bears directly on the healthy welfare of all creatures. Indeed, many city dwelling humans anonymously cramped in office cubicles and tiny apartments are the ones with no reciprocal relationships. Does that mean city people have no moral standing? What about hunter-gatherers and other indigenous people who reciprocate with their environments through others, including animals. Do they count less than the money makers crowded on Wall Street?

Korsgaard believes a case for human obligations toward animals can be made; but it's dubious which value exactly she notes, citing how Kant bestows higher approbation on those possessing rationality. By virtue of their evolved intelligence and social systems, great apes have value for and offer value to ecosystems that benefit others, asking little in return. Korsgaard does not absolutely exclude "animals" from thinking about what is good for themselves, as Kant who sees only a rational being as lawmaker of what is good for others. Yet she clearly intimates that animals don't go the next step and create rules for others about what is good. Here again, as elsewhere in her book, Korsgaard reveals the huge hole of not discussing primates or apes, but for a passing reference occasionally. Rather, she claims the law is on human shoulders to grant moral standing to animals. Is this because she believes humans are superior? Granting moral standing is important, but it only and continually elevates the "human" over the "animal" and gives the human sole evaluative power over the self-interests of all other organisms. As I hope to demonstrate throughout my book, and especially in chapter 3, apes are indeed moral makers in their own right.

Some might not technically see Korsgaard (2018) as an objection to my claims. Thomas Nagel (2019), for instance, emphasizes the pro-animal thrust of Korsgaard's argument. Broad strokes about "animal rights," however, miscalculate gradations of degrees among species. Nagel, summarizing Korsgaard, suggests that "other animals" live only passively in the present. Both thinkers ignore the field research and lab data, mounted and discussed in my book, showing that great apes actually think fairly similarly to humans. Apes actively self-assess. Apes are neither physically, psychologically, nor morally passive. This is why, meeting skeptics, I focus on apes and don't generalize about various "animal" species. One can't compare apples and oranges; rather, one must compare (and contrast) what they have in common, as I've done in this book between humans and apes (and even among great apes) with the critical difference of apes as ethical ecosystem engineers.

In any case, great apes should not be acknowledged by humans as a valuable species only because they are eco-engineers but because they are moral individuals determining their own good outcomes in a common unity of others like them. As Korsgaard (2018) herself suggests, many animals have a *temporal unity* of mind. I simply aver that with great apes that temporal unity extends to the past and future with horticultural actions involved. For a critically endangered species, against a utilitarian like Gary Varner (2012), each individual expressly counts, as even Korsgaard might say. However, ape species as a whole contribute constructively to necessary biodiversity and tropical forest ecological health on a large scale. A forest does not consist of one leaf, but of leaves on a branch, as part of a sapling, in a system of many trees, evidenced in the notion of a mother tree communicating invisibly with younger trees through its roots. Thus, somewhat in distinction to Korsgaard, I do place value on ape species by granting them moral standing, since they engage in forest health through temporally social and foraging unity as moral individuals. Where's my proof? Look to the forests.

I hope this section has helped to shed light on my particular argument and claims. Before continuing with a discussion of ape mentality in chapter 2, let me conclude this chapter with some final words about ape morality.

DO APES HAVE MORAL MINDS?

This is not an easy question to answer, though some guideposts have been offered so far and there will be a more full discussion that follows. For this section, I will draw on work by animal biologist and conservationist Marc Bekoff and bioethical philosopher Jessica Pierce (2009) in their apropos book, *Wild Justice: The Moral Lives of Animals*. It's already well documented that rats, monkeys, cats, bats, dogs, and elephants, to name a few species, exhibit varieties of "moral intelligence" in how they express compassion, empathy, altruism, and punishment to their own and even other species. Developmental psychologist Paul Bloom (2010) has written about the moral life of babies who not only have an early mental existence but who can sense another's pain. All of this is especially true of great apes. Are philosophers to grant "morality" only to humans, then? If so, that is speciesism in light of research that shows the biological evolution of morality (Tague, 2016, chapter 2) and the moral systems in great apes (de Waal, 1982). Bekoff and Pierce go as far as saying that during interactions animals know the difference between right and wrong, evident in a range of social emotions that include trust and forgiveness. Confidence in and empathy for others might be hardwired, they say, since these responses are hormonally related to oxytocin across mammals.

For apes, a cognitive emotion like trust is paramount in relationships with others since, for the most part and even with hierarchical systems, they strive for social equilibrium and not outright chaos. Bekoff and Pierce (2009) say there is a direct correlation between social and moral complexity and a further link from there to honest and reciprocal cooperation, empathy, and justice. I don't rely on the concept of "justice" in my book and might point to, instead, how great apes *as individuals* morally comprehend equity, fairness, and forgiveness. Equally important are those instances where, even over a period of time, an ape remembers a harm or injustice done to him by a conspecific or a human and later, given the opportunity, seeks a remedy. Emphasis here is on showing that apes, via moral individualism, extend moral behavior not just to social but to ecological environments. Apes seem to grasp that their social spheres inhere in a natural habitat that requires care and maintenance. Can I unquestionably prove that? No, but it's obvious on its face that they have not wreaked havoc, and the empirical evidence I provide points to an ape societal-ecological homeostasis.

Ape social intelligence extends to their land ethic. Where Bekoff and Pierce (2009) broadly relate "morality" across species, moral individualism is applied here empirically, meaning that each ape's actions, socially and environmentally, contributes to the ecological coherence of her group and habitat. Where they look for broad patterns across mammal species I focus attention on hominids and communal, including ecological, behaviors. As Jane Goodall (1986) notes, most chimpanzee interactions are nonaggressive and could even be termed *peaceful*, a prosocial life not mired in physical conflict. As Bekoff and Pierce say, primate research shows that most relations are affiliative and not agonistic. Much animal activity, in fact, is involved in play, grooming, social exchange, or quiet individual or company foraging. These are expected behaviors controlled by group norms. Underlying these standards of conduct or *mores* are genes and then learning. Just as prosocial behavior is not an exaptation but governed by natural selection, so too is the community oriented land ethic of apes. Just as social beings conform to avoid sanctions, on some level conscious or otherwise, apes have adapted themselves to conform to and not to be in conflict with their ecological habitats. Following Tinbergen's framework again, here's what constitutes the ape land ethic as demonstrated in the two main chapters that follow.

1. Coevolved with social behaviors like foraging;
2. Proved an adaptation for fitness;
3. Was and continues to be caused by some understanding of the link between personal/physical and community/ecological homeostasis;
4. Is a product of a long phylogenetic and ontogenetic development.

As scientists routinely make inferences, I do as well. There is talk of continuities, but I realize that continuity does not mean there are absolute equivalencies. Cross-species comparisons can lead to misunderstanding, but given the contours of speciation from common descent they are useful. Some might accuse me of anthropomorphizing. How could that accusation be true when decades of lab research and fieldwork reveal that apes have intelligence (e.g., problem solving and tool use), moral emotions (e.g., fear, jealousy, happiness), and social norms (e.g., approval and disapproval) similar to those in human beings? Frans de Waal (2019) says that considering the close human evolutionary history to apes, it's "logical" for humans to think anthropomorphically about them. As Howard Gardner (2011) has pointed out, there are multiple intelligences, and they vary across species. For instance, the spatial intelligence of an orangutan enables her to amble gracefully across a rainforest tree canopy. Yet there is similarity in social intelligence between and among apes and humans, for how else is one to keep track of the transactional maneuvers of another.

Apes not only cooperate socially but reciprocally service their physical environments. That feature of including one's fitness in the environment is what sharply distinguishes primates and especially great apes from most modern human societies. This is difficult to explain but gets to the heart of ecosystem engineering. The physical action of primates as they move, clear branches, and eat in a non-ravaging manner serves them and, at the same time, promotes forest vibrancy. At the basis of moral behavior is the ability to be conscious of another's needs and desires. In a related way, apes appear to be aware in how they economically regard their habitats. Just as moral codes pivot on cooperation, a land ethic too, means that individuals and groups cooperate with and do not ravage their ecosystems. As far as is known, ape species have not ruined habitats (shrinking as they are) but have, on the contrary, acted in a mutually conducive way to help habitations flourish. Just as in moral behaviors one finds reciprocity and sharing, we can say that apes are motivated to act reciprocally with their biotic communities. Genetic relatedness causes kin-to-kin favoritism, and while this can be extended to non-kin, similarly apes treat their environment in a kin-like manner.

There are long-term benefits to social and environmental cooperation over any immediate gains. Apes appear to have evolved in their biomes mutually with other plant and animal species. Therefore, some of the cornerstone emotions for moral behavior, like empathy, sympathy, and reciprocity could be at work in what is called the ape ethic toward their land. How can one know? Flip the table and see, for many conservationists will ask humans to act in such a way: be kind to the environment. None of this has to be cognitively driven. The evidence at hand indicates that primate species, not counting humans, show at least an unconscious, instinctual, adaptive regard for their physical surroundings, as do traditional hunter-gatherer societies.

The oldest part of the mammalian brain deals with survival. Evolved onto that is a middle brain that emits emotions. A further adaptation, all segments working in unison, is the higher level brain with cortices that govern thought and emotional response. Humans share this interconnected brain topology with apes, and most people comprehend that if they destroy their environment their survival is at risk. Ape and human brains share mirror neurons and spindle cells, meaning one sees oneself in another and can feel another's pain, generating moral agency. This agency, I say, is applied to the ecology of ape habitats as well. Apes might see themselves functioning as intimate cooperators with nature. Bekoff and Pierce (2009) talk about, in terms of social relationships, the notion of an *expanded self.* Apply this to the moral individualism of apes in how the ape ethic invests an emotional empathy in the physical environment. There's some truth in my thinking, evidenced in how most modern humans allow technology to remove them from the wilderness resulting in no emotional contagion with green surroundings and, therefore, no land ethic.

Just as there is a moral instinct for apes, governing approach/withdrawal or approval/disapproval, there is a land ethic. Impulse control for all ape species, including humans, is a higher cognitive function and implied in the outcomes of moral instincts. However, for many humans, the impulse to control disregard of land is not often regulated, as it seems to be among apes. This is not to say that great apes are not proximally without substantial appetites, but they have ultimately managed to evolve themselves (their life history) to act according to Aristotle's rule of middle state moderation. They are just and temperate in their wide-spaced feeding, varied food choices, and through ecologically secure population levels. Ape passions about consumption are not irrational, as are human desires. For example, how many brands of any type of food product do we need? Why have we chosen to eat every bit of many animals, sometimes while a creature is still alive, down to stomach enzymes? In terms of virtue ethics, many humans are the ones who are deficient in their omnivorous intemperance. Some may say this statement, along with others I make, is editorial exaggeration. Perhaps, but readers should look at the facts and determine who among hominids is the paragon of consumerist excess and temperance deficiency.

Because humans are supposedly smarter, more moral, and even "spiritual" over animals, they should know better not to recklessly disadvantage apes and their habitats. As Tom Regan (2004) says, a moral theory is required to take into consideration obligations from others. Of course he's speaking about how humans ought to treat animals with respect because of their inherent value. Another way of looking at this, though, is how great apes treat non-kin with care or respect (e.g., reconciliation, consolation, or avoidance) and similarly show regard for their physical environment, where there is no wholesale destruction. These are evidence-based duties and responsibilities of assistance and restraint that enable great apes to live socially and environmentally sustainable lives.

In making the case for an ape land ethic thus far, I've covered in this chapter some overarching issues in preparation for the two thick empirical chapters that follow. Therefore, readers need to keep in mind seminal ideas, problems, and questions just covered, such as forest personhood, moral individualism, ecosystem engineering, individual agency and autonomy, the ape moral ethos, and objections stemming from the long shadow of Kant's focus on rationality, which is in sharp contrast to my emphasis on ape intelligence evidenced in their bodily consciousness.

Chapter 2

Cognition and Intelligence in Environmental Adaptation

This chapter, linking back and looking forward, tries to show how apes use their intelligence as moral individuals in a community and forest ecosystem. In fact, intelligence in part may have arisen from environmental and not just social pressures. Cognition and intelligence as presented here, granted, lean on mental representations and inference, but I hope readers will see that the thrust of the chapter is more to aspects of intelligence via cognitive embodiment, which I address at the beginning and then swing back to at the end. The danger of presenting some of the current literature on ape mentality is that inherent in it is a lurking Cartesian duality where the subject is removed from the world. But I don't believe that's so, and I hope readers see the gist of the book, in discussing such views, is really to show apes as thinking eco-engineers with a land ethic that situates them as inseparable from their environments. However, ape intentionality is embodied and relational and does not, essentially, require any representations that put them outside of the world looking in. Nevertheless, I try to square computational and embodied perspectives to demonstrate how apes are capable of intellectual inferences that enable the completion of mental puzzles for social and ecological needs and desires. Apes are neither abstract philosophers nor are they instinct-driven robots. They are smart and use their forest intelligence in ways somewhat alien to us. Since my argument is cumulative, some of this discussion and these claims will spill into chapter 3.

While this chapter offers an overall discussion of ape intelligence, readers are asked to keep in mind the true focus of this book: that is, how intelligence relates to moral individualism and is used socially in eco-engineering. I will bring up those focal points as needed, and this is probably an opportune time to mention work by philosopher Shaun Gallagher (2009) who informs us that much of modern philosophy has ignored the body and its place in the world by eyeing only the mind. In contrast to a rules-based Kant, Gallagher considers Aristotle, who proffers *phronesis*, knowing how to act in a given context. Are great apes normative creatures? I believe so, even regarding exchanges with their busy habitats, and will continue to provide examples. Certainly cognition is not narrowly individualistic nor is it passively ideational, Gallagher states. Rather, cognitive experience is an organism's reaction among transactions in an environment. Likewise, Edwin Hutchins (2008) speaks in terms of cognitive systems as distributed across places, objects, and events, transcending limits of the brain itself. Thinking is not just a consequence of brain activity but represents body/brain interaction with ecology, seen in cultural and social practices. This would be especially true for apes whose culture is ecologically based. Who more than great apes help us understand that, in spite of their advanced mental capacities and abilities, the physical world does not just happen in their heads but through their bodies among others in ecological circumstances, a situational cognition.

Gallagher (2014a) affirms what he calls *embodied cognition*, or how in one's perception of the world there is simultaneously a keen awareness of one's moving body. Embodied cognition is, clearly, more true for apes than for people who work at desks. Objects in this ape world are possibilities, Gallagher might suggest, perceived through an entire corporeal form particularly situated. Consider these ideas, I'm asking, in what you read here and the next chapter. In this sense, objects are not just fruit trees or tools to access honey or seeds but are other apes using such items and thus creating an intersubjective embodiment along with the forest itself. This heightened awareness implies agency, meaning that an ape's body is not merely an object among other things but, as Gallagher (2014a) would say, "affectively rich" and filled with emotions, desires, and needs. These affective states operate in a dense ecology and impact perception and hence cognition. The point I'm making through Gallagher (2009, 2014a) is that great apes are not just thinking mechanisms with drives to survive but cognitively aware, intelligent beings shaping and shaped by the environs in which they are intimately active with their kin, peers, and other forms of life.

Though related, the terms "cognition" and "intelligence" are neither synonymous nor interchangeable. Cognition would be the underlying mechanisms in the process of mentality. Intelligence is the ability to take the elements of that process and apply them to different situations in order to solve problems or make decisions. Cognition is evolved in brains; intelligence is output, sometimes social, and a completion of cognitive activity. Richard Byrne (1995) sees intelligence as a plastic means to overcome ecological or social challenges and effect change outside of genetic (cognitive) constraints. Intelligence, as Jane Goodall (1986) echoes, is the ability to vary knowledge and responses to objects under different circumstances. Intelligent behavior is flexible and adaptive. Capabilities tied to intelligence include learning by association, imitation, and insight for novel behavior. Developmentally, the ability to imitate suggests mental representation and symbolic capacity because it involves interior thinking that is manipulated as external action (Miles et al., 1996). Intelligence among apes enables them to refrain from instinctual action so as to consider alternatives. Nevertheless, the word "intelligence" is used with some caution, since there really is no hierarchy among intelligent creatures. This is because each organism evolved cognitive abilities in different environments. No small matter, since moral individualism, as part of the ape ethic, depends on how a primate's brain adapted to its physical and social niches.

Regarding moral individuality, it is often the narrow perception of many humans which blinds them to the sentience and sapience of ape persons and encourages seeing them as "a set of capacities," according to philosopher Lori Gruen (2019). This is, in some part, the danger of stressing ape cognition and sociality. Great apes are decidedly intelligent and sociable but should not be reduced to those functions and abilities. Rather, they employ their evolved cognition and social behaviors in many morally individualistic and ecologically collective ways. Readers are reminded to keep in mind that as the portrait of ape personhood is drawn on this canvas, all the particles from all chapters are necessary to create the whole, not a single stroke of the brush from any one chapter.

EMOTIONAL UNDERSTANDING

According to Kristin Andrews (2015), cognition mediates sensations, memory, communication, reason, and behavior. A cognitive process itself can cause behavior. Andrews tells the story of the zoo chimpanzee Santino who would regularly stockpile hidden stones to throw at visitors. With this evidence, Andrews asks, how could skeptics deny ape cognition regarding forethought. Kant, on the other hand, believed animals are driven by desires without understanding. Some might say Santino merely associates different actions and objects without any underlying cognitive mechanism. There seems to be a huge gap in discovering outcomes between those cognitive theorists studying the minds and behaviors of lab primates and those hardy researchers conducting fieldwork. What an ape inherits biologically comes in large measure from the biome in which she evolved. Since genes respond to environment and learning, it is logical to ask: What is the cause and purpose of a behavior, or, how does an ape function in her habitat?

Yuval Harari (2017) asserts animal intelligence and decision making because natural selection has weeded out, in his words, timidity and rashness to favor calculation. When people stopped visiting Santino for fear of getting hit with a rock, the chimpanzee began stacking sleeping hay at the front of his enclosure to hide stones from the view of unwitting passersby. In time, Santino devised other hiding places, like tree trunks. Skeptics refuse to assign complex consciousness to Santino's actions asserting, instead, that his movements typify the evolutionary algorithm or method of calculation, according to Harari, of a male chimpanzee. I say, however, given what we know of the mentality of apes, that it's blindness to assign Santino, surrounded by human activity, to a one dimensional reality, that is, that of a mechanized chimpanzee stuck in some evolutionary past. Rather, Santino is an example of the making of mind or how an individual brain (self) deals with circumstances outside of his true habitat.

Even among apes there appears to be moral sense and individual consciousness. Darwin (1859, chapter 11) says there's no law of necessary development. This means that an organism's variability and evolution are more dependent on relations to other organisms than to the physical environment alone. While this observation rings true, clearly, however, much has been learned since Darwin's time about the codependent and coevolutionary interactions between organisms and their environment. For the purposes of this chapter, though, emphasis for the most part is placed on the social environment, initially, with a look at environmental psychology later. The point is that great ape cognition is dependent not only on their underlying cognitive mechanisms but also on their long curve of learning how to deal with others in a forest domain (or in the case of Santino, an artificial, prison-like world).

Since the physical environment for apes includes communities and groups of conspecifics, there's a strong social component to cognition and intelligence. This is evident even of the semi-solitary orangutan who interacts with others closely or at a distance. Take communication in apes. There are signals and messages in gestures, vocalizations, and eye contact critical to cooperation and learning. There is intentional communication in apes in a defined context, and so they emotionally and intellectually understand another's needs and desires. Mind reading is part of social communication and, in turn, part of cognition and intelligence. There is more, however, than mere learning, as important as that is. I might hypothesize there's an ape emotional comprehension of learning that coordinates not only among and across species but with the forest environs, what might be referred to as a transfer index (Savage-Rumbaugh, 2018) or how an ape person gauges and verifies her own intelligent learning in terms of eco-engineering.

The function of cognition is both controlled and automatic, where there is a dynamic exchange between higher order thinking, direct sensation, and memory, all of which aid in object and event recognition. While elements of cognition are learned, and hence the long interval of infant ape development under the tutelage of the mother, there's nevertheless a conceptual component of classification or inference to knowledge used in understanding the world. Concepts linked to memory and associations form thinking. An ape might remember how another individual acts in a certain environment so as to conceptualize behavior for others similarly placed. Perception, therefore, is a key part of the cognitive process, and apes are keen perceivers with excellent memories who can associate objects, events, places, and others. Borrowing from nineteenth-century cultural anthropologist E. B. Tylor, there is a psychological wholeness for ape species who carry in collective brains the accumulated evolution of their behavior. Indeed, there's a rational comprehension in grief. Consider how an ape's intense emotional response is measurable in body language and movement on the loss of a mother, infant, or companion. This picture shows that there's a fine line between feeling and understanding (King, 2013). I speculate that this connection between feeling and intelligence tempers ape involvement in their bio-networked environs, too.

Drawing on some of the researchers cited here and in my next chapter, Steven Wise (2002) offers a comprehensive summary of the remarkable mental abilities of the signing orangutan Chantek and the signing gorilla Koko, such as mental representation, logic and math, self-awareness and intention, joint attention, imitation, deception and pretense, theory of mind, and autonomous self-sufficiency. Steven Wise argues for the legal rights of great apes. However, the difference with the line of thinking in my book lies in the claim of an ape land ethic as moral individuals because of their ecological engineering, explained in chapter 1. What does theory of mind or moral individualism have to do with eco-engineering? The conscious intent of ecosystem engineering does not have to be proved. There are analogous instances in the human realm, like the instinctual and learned protection of and care for a home base. We know apes, especially orangutans, are extremely intelligent. Whereas chimpanzees react quickly, orangutans thoughtfully ponder before responding. It's not enough to catalog ape skills. We need a theory that explains why they have not completely whittled away their diminishing resources (in spite of human intervention and intrusion) and why they have not destroyed each other. More so, great ape intelligence and social behaviors evolved along with their abilities to steward resources and to acknowledge the uniqueness of another's personality, needs, and motivations. Compared with humans, the ape life history of few births at long intervals, intimate contact with nature, and so on, meshes constructively with the web of their forest homes.

Sue Taylor Parker and Michael McKinney (1999) outline the various theories on the evolution of intelligence.

1. The social brain, consisting of examples like chimpanzee politics and Machiavellian intelligence. This perspective includes reciprocity,

account taking, competition (especially in groups), and risk assessment to body or alliances.

2. Ecological, which deals with resource exploitation. Fruit eaters have greater comparative brain size than leaf eaters (*pace* gorillas) among primates. This alternative to the social brain hypothesis includes food competition across space and over time. One needs to know what to forage for, where, and when. Richard Byrne (1995) says the salient selection pressure for ape insight was overcoming environmental obstacles, whether in manipulating resources or tools, teaching and learning about resources, physical movements in space, or hunting. Gen Yamakoshi (2004), too, asks if the evolution of complex feeding techniques is the origin of ape intelligence.
3. Socialization, a combination of the Machiavellian and ecological. Social and food resourcing skills work together, and these include imitation and teaching.
4. Locomotion. Self-awareness comes via tree climbing, movement, and hanging, especially in orangutans (Povinelli & Cant, 1995) who will simultaneously suspend themselves while foraging and eating. Here there's a strong mental representation of the body in motion and all the limbs and digits working in a synchronized strategy. Kevin Hunt (2004), too, says that the special demands of great ape locomotion and posture contributed to the evolution of their intelligence.

Anne Russon (2004a) tempers the above outline as follows. The social brain hypothesis does not consider complex ecological food sourcing and instead places too much emphasis on group size ignoring any hierarchical complexity and important small scale social structures or individual inputs. Cognition is not a constraint on sociality. Concerning the ecological hypothesis, great apes are not alone in this type of food processing. Russon adds what's called extractive foraging, which is seasonal and requires tools. Here, too, though, this is not limited to great apes and appears, for example, in baboons. Tool use does not appear in all chimpanzee or orangutan populations, however. Russon also talks about apprenticeship, or how extractive foraging needs a long social development. Like socialization in the above paragraph, this approach integrates social skills and ecological knowledge. In terms of locomotion, sometimes referred to as arboreal clambering, Russon notes how evolutionarily older orangutans, and even living ones in Borneo, are prone to terrestrial movement.

Lower scale family units appear as part of larger social structures in humans and gorillas, according to Robin Morrison et al. (2019). Somewhat negating the social brain hypothesis, their research suggests that, as with humans, kin relationships are responsible for hierarchical group memberships in connected tiers and networks, not loosely scattered collectives. For me, minimizing the social brain hypothesis (though not completely eliminating aspects of it) seems to shift emphasis to eco-psychology, embodied cognition, and ecosystem engineering, whether in territoriality (chimpanzees) or overlapping home ranges (gorillas). Especially for gorillas, there are mutual gatherings at key mineral and protein rich resource locations with some affiliative interactions between groups. These symbiotic, village-like group relations could have been influenced by, in addition to standard reference of the mother-infant dyad, the long and close association individuals and assemblies have to forest systems which act as ape settlements. As with humans, repeat and reciprocal interactions among various gorilla clusters reflect the plasticity of hominid evolution in the context of high quality resource foraging.

From some combination of these theories (numbers 1 to 4 above) great apes have evolved cognition and intelligence superior to that seen in monkeys and in line with what develops in *Homo*. With the ape land ethic as moral individuals because of sustainable ecological engineering, the emphasis here is on social intelligence *and* premeditated foraging, particularly extractive, especially where "social" includes collaboration with the forest environs.

Social interactions among apes, including humans, are probably more emotional than intellectual. Ape "language" is less a matter of mere signaling and more to coordinated, regulated, socially communicated meaning, as Barbara King (2004) might suggest. No solitary movement or facial expression belies a single emotion. There are many embodied, contextual parts between individuals that constitute sociality. Noteworthy patterns of behavior emerge from qualitative descriptions and not quantitative facts, evidenced in great ape field narratives by Jane Goodall (1971, 1990), Dian Fossey (1983), and Biruté Galdikas (1995b). Communicative meaning is constructed through defined and decisive gestures. Perhaps oversimplifying him a bit, Gordon Hewes (1973) relates how early hominin australopithecine proto-"language" was gestural, including mouth movement, and only gradually evolved to become vocal where some gestures and vocalizations mixed. The point, following King, is that even among apes communication is regulated contextual meaning in social interaction patterns based on individuality, group construction, and ecological environments.

BRAIN AND CONSCIOUSNESS

As neuroscientists Michael Graziano and Taylor Webb (2014) could ask, what's the adaptive value of consciousness? Their answer focuses on what's called the attention schema theory. That is, consciousness came into being since an overload of information stimuli had to be managed, first on a neuronal level, later on a basic cognitive level. The eye is a result, over millions of years of evolution, of this process, as is the brain, which started as a tectum over 500mya to filter incoming auditory and visual signals so as to make survival or food resourcing decisions. Over time neurons and their networks evolved together so that brain tissue could experience not only overt attention, or what's happening out there, but also covert attention, or what's happening in one's mind. Moreover, as the psyche of various primates evolved, the ability to guess at another's intentions developed. Consciousness, then, is elemental to higher order thinking and separates species more in degree than kind, where the consciousness of great apes is likely similar to that of humans, though that assertion is hotly contested as will be seen in my upcoming section on theory of mind. For this discussion, attention and consciousness are related to the body as a tool, or how an ape attends to her habitat through her physical awareness of self and others including acute mindfulness of the ecosystem.

Concerning the question of multifaceted conscious awareness, which some researchers (upcoming in this discussion) are fond of minimizing in apes, consider chimpanzee patrols of their territory. This behavior is not only deliberate and coordinated, but it's performed in near silence so that one can detect any signs of the presence of a stranger via sound, sight, smell, feces, urine, or food leftovers. Cannot behavior evolve to the point of complexity where conscious decision is possible? As another example, an orangutan will deliberate about the best route for food resources based on any number of factors, from memory and weather to the spatial location of others. A scientist who denies intricate awareness in an ape, meanwhile using the expression "like us," might then make analogies to activity or time budgets of humans. Scientists will say that any ape "strategy" is not conscious but, rather, an evolved feature that provides a benefit. While this might be true, generally, in "mating strategies" (even for humans) is there no choice at all in, for example, a preferred mate or a consortship?

Philosopher of animal ethics Zipporah Weisberg (2015) points out how Tom Regan (2004) does not see rational agency in animals (via Kant) but sees them as "patients" who deserve moral concern. Then, contrary to Regan and on the other end of the spectrum, some self-styled post-humanists, says Weisberg, grant agency to all living organisms, beginning with bacterial microbes. The middle ground, with a focus on apes, is a good place to start since animal agency is, for most people, a radical idea. Consciousness implies agency, and many animals are complexly conscious and act intentionally to construct their own meaningful world, an *Umwelt*, by interacting with other animate agents and inanimate objects.

To some extent, we can say that the brain of each species equates to its identity. Katerina Semendeferi (1999) quotes early researcher C. J. Conolly as saying that the orangutan brain because of its shape, high frontal lobe, deep orbital region, and frontal lobe variability makes it among ape brains the "most interesting." The orangutan brain is more similar to a gorilla's than to a chimpanzee's. While the prefrontal cortex brain fissures in humans and great apes are very similar, says Semendeferi, the cingulate sulcus (part of the limbic system dealing with emotions) among humans, orangutans, and gorillas is distinctive. This scientist found the frontal lobe in its entirety is larger in orangutans and chimpanzees than in gorillas, but the orangutan orbital sector associated with sociality, unsurprisingly, is smaller. Different selection pressures play a role in brain morphology, but brain volume and body size do not reveal adaptations in neural circuitry. This brain, surely, has become a vital engine in forest adaptation away from the human brain.

André M. M. Sousa et al. (2017) studied the molecular and cellular reorganization of neural circuits in ape and human brains. All regions of the human brain have similar counterparts in the brains of apes. Though there is some distinctive gene activity within human brain regions, the human brain is simply a "larger version" of a primate brain. There are similarities across ape/human regions, notably and perhaps not surprisingly, in the prefrontal cortex, responsible for sophisticated cognitive and social behavior. This similarity would, of course, be important for my book's line of thought, evident from the ideas laid out in chapter 1. The striatum, crucial for voluntary movement but implicated in social behavior, has the most human-specific genes. Putting aside some other similarities/differences discussed by Sousa, of note is hot tyrosine hydroxylase (or TH) for dopamine production (a neurotransmitter involved in reward and reinforcement) in higher order thinking. This chemical is highly expressed in the human neocortex and striatum but not existent in chimpanzees. The TH gene was probably lost in our shared evolution and then reappeared in humans. Cognitively, however, there are similarities between the human and ape brain. Is a brain, however, a mind? In answering, one needs to consider how consciousness consists not only of brain matter but also of culture and physical environment, indicating a major difference between ape and human ecological motivations.

The question is where, and if at all, any line of demarcation should be drawn between human and animal minds, according to Gordon Gallup (1985). Since Gallup's paper was published, much research has been conducted pointing to animal consciousness (e.g., Feinberg & Mallat, 2016). Animals can become objects of their own attention, evident from mirror self-recognition. This is important since self-awareness permits one to estimate another's thinking and make inferences from one's own frame of mind. Self-recognition is central to socialization, for it creates a sense of mental common ground between self and other, the modeling of a virtual visual field of another outside one's own. Following John Locke, Mark Rowlands (2019) sees awareness of self and other as a pinnacle of personhood. Recognizing one's own thoughts in another person, and, especially how one's thoughts often differ from another's, thinking into another mind, is sophisticated on an evolutionary scale. As my book attempts to demonstrate, research shows that apes (with some variation among species) understand another's intentions, assume others are responsible for actions, act as if to anticipate and influence another's behavior, and comfort others and reconcile transgressions. These are all evidence of the existence of mind, Gallup would argue. If mind does not constitute personhood, what does?

Can apes form a mental sense of awareness about the forest bio-network? Apes have concepts since they sense, for example, the presence of a fruit tree, over there, and then get closer to inspect which variety of tree and whether it is a desired food resource. Based on their social and ecological behaviors, there's good reason to believe that apes, like humans and other animals, form concepts from prototypes both experienced and genetically housed. While an ape is predisposed to feed on trees, she must learn which ones supply the best nutrients and which can be used as a second source. These concept schemas are vital in the life of an ape so that once mature she can devote attention as an individual to other matters and form new and more complex mental representations.

Contrary to Plato, knowledge from the senses, especially for an ape, is not wrong; in line with Plato, apes have conceptual ideas about, for instance, where to habituate themselves. As with humans, ape thinking is not arbitrarily plastic since there are innate forms and instincts. Konrad Lorenz (1965) notes how learning is an evolved disposition with an adaptive outcome, so there are automatic (e.g., calculation) and controlled (e.g., skill) processes. Unmistakable from the experiences of field researchers, finally, one can argue that in line with nineteenth-century philosopher and psychologist William James, apes experience a stream of consciousness that is both within and outside of awareness. This consciousness stream for apes, though, is geared in large part to the ecosystem. As revealed from long-term observations (e.g., Goodall, Fossey, Galdikas), over the course of development and as part of life history, each ape builds her or his own mind, socially and ecologically.

In "What Is It Like to Be a Bat?" philosopher Thomas Nagel (1979) suggests that one measure of an organism's conscious state is not to look at its intelligence but its subjective experience. Intelligence is not equivalent to sensory consciousness since all organisms experience the same environment differently. How a bat maneuvers at twilight using echolocation to identify food is vastly different from how an orangutan slowly crosses a tree canopy in the morning in search of sustenance using her vision and memory maps. Nevertheless, treetop movements by large-bodied species like orangutans could be responsible for evolving a psychological mechanism related to the conception of self (Povinelli & Cant, 1995). Cognition has almost nothing to do with how an organism feels, and affective responses are probably very old in evolutionary time. Consciousness is an embodied process of brain function, an internal world of affective states and cognition. Consciousness, whether through smell or sight, is a way of processing, interpreting, and remembering sensations. Of course consciousness is social intelligence, as it includes variables such as feelings, emotions, sensations, and attention.

One key implication, I posit, is that in consciousness apes "see" their environs along lines of inclusive fitness, as if the habitat is a related being. This analogy postulates that apes intuitively consider the forest ecosystem as an entity for whom they are responsible and on whom they depend for genetic longevity. Is this bodily consciousness a cognitive ability? So ask Liangtang Chang et al. (2017). In tests conducted on rhesus monkeys by Chang, mirror self-recognition was evident after learning physiological responses to stimuli associated with mirror images. This study is important for several reasons. First, it shows how a concept of self can be learned. Second, recognizing the self in relation to others permits symbolic consciousness where the internal world is expressed externally. If this is so in monkeys, magnify the results and implications in great apes. Importantly, failing a human-constructed mirror self-recognition test does not equate to an animal's lack of self-awareness. Seasoned researchers like Nicholas Mulcahy and Josep Call (2009) caution that tests on apes designed by humans can produce false outcomes. For instance, in this case, Mulcahy and Call discovered that in an object-choice task test, unsurprisingly, distal and not proximal location by the ape ensures better cue comprehension. That explanation helps us understand ape behavior in a forest environment.

Human researchers cannot assume to know how apes will act or what constitutes their consciousness. Nevertheless, Tom Regan (2004) argues that animals can have beliefs tied to desires. Beliefs do not necessarily require concepts that depend on language (Andrews, 2016). One believes something about what is desired or believes one wants something one lacks. When a female orangutan sees a male moving in the trees, she truly believes he is a threat or a potential mate. Does an orangutan's belief, based on experience and context, about certain foods grant her a taste humans don't have? Humans cannot experience an orangutan's world the way she does. Can one reasonably imagine what the orangutan feels in certain circumstances? Yes, for she feels pain, suffers, and yet experiences contentment like a human person. On the other hand, there are differences in terms of ape/human environmental behavior as I am showing.

Sue Savage-Rumbaugh et al. (2000) see consciousness as a property the brain manipulates. For instance, because chimpanzees are good imitators, copied behaviors can pass on or die out if not imitated; in and of themselves they are not inherited per se. Culture builds off a small, evolutionary basis. Culture is a means for an organism to adapt in ways biology might not permit. So for ape species, not only humans, consciousness and culture have coevolved to help form tools, systems of communication, complex social organization, and environmental responsiveness. Because apes have culture, we know, then, that they have advanced consciousness. However, the ape ethic is not equivalent to ours since they evolved to fill different types of eco-centered social niches.

So it is a mistake for the general public to cite what they see as deficiencies in apes when comparing them to humans. There are, rather, strengths and continuities with salient differences. The bonobo Kanzi, Savage-Rumbaugh (2000) reports, showed a remarkable ability to absorb, process, and express the human culture and language in which he was surrounded. Kanzi was able to understand and construct abstract meaning not only with his human caretakers but with a wild bonobo with whom he was raised, demonstrating a transfer of linguistic culture. Patricia Greenfield and Savage-Rumbaugh (1990) say that Kanzi could "create" a sentence. Their research shows how an ape can invent grammatical rules. This implies that what we call grammar might not be a human-only mutation. The language capacity probably comes from a common ancestor who sequenced sounds and representations for action. Seasoned researchers like Savage-Rumbaugh et al. (2000) firmly believe that an ape brain and consciousness can develop differently if raised in a human culture of spoken words and symbolic forms. The implication of that assertion about great ape neural plasticity bears analogously on my ape ethic as eco-engineers since it implies how these primates evolved as forest persons to socially communicate with their lands for mutual preservation.

What was the primate mind before material culture (e.g., tools), ask Robert Seyfarth and Dorothy Cheney (2013). From a social perspective, and evolutionarily old, primates hold concepts of individuals in mind. This idea of individual identity, say Seyfarth and Cheney, means that primates can negotiate an ensemble of traits to reduce a conspecific to a representational type. In this way, a baboon, for example, can reasonably expect an individual to act in a certain way, under certain circumstances, and so creates a theory about that other's behavior. Taken to another level, I conjecture that the ape mind creates a theory about the environs. Classification of personalities, or as I'm suggesting, particulars in habitats, is a higher order thinking skill not evident in human infants, and yet this innate ability paves the way for later adaptations of mind during development. Richard Byrne (2000) believes that the *Australopithecus* and chimpanzee split was nearer to 4mya and not 6mya, meaning that the early appearance of hominid mental capacities is much older since adaptations evolve slowly. What are behavioral instances of an ape environmental theory? Savanna chimpanzees use tools to extricate underground root organs in behavior likely similar to early *Homo* as well as *Australopithecus* and *Paranthropus* (Motes-Rodrigo et al., 2019). Jennifer Vonk (2014) says that orangutans imagine higher quantity as a mental representation better than other apes. In an earlier study, Josep Call (2000) shows how orangutans employ a representation from memory to compare and discriminate among quantities not readily visible, which is useful in foraging food portions in different locations.

Not as far back as 30mya when the primate brain appeared, but surely by 12mya, primates and thus the descendants of modern apes, says Byrne (2000), were able to form mental representations and grasp consequences, cause, and effect. Over time and because of natural selection these capacities would develop in complexity across ape lineages. Such cognitive advance is probably from an increase in brain size but especially brain organization. Nonetheless, Lauren Gonzales et al. (2015) have shown that there was brain complexity before encephalization in old world monkeys of 15mya. However, sensitivity about ecological tool use among apes is a significant advance and distinguishes them from monkeys. Byrne goes on to say that chimpanzees and orangutans know the difference between an accident and an intentional act. Byrne suggests, then, that great apes, over monkeys, have a super ability when it comes to forming and manipulating mental abstractions. Could such "thinking" be part of their evolved eco-engineering? In contrast to monkeys, apes seem to know why someone does something. Is fundamental human intelligence so different from that in great apes? Classification is part of animal nature, as in discriminating between food and non-food sources. In other words, there is more than mere perception. Furthermore, abstractions about food, for instance, can overlap in features and become conceptual in relation to other species, as in the notion of eating (Vonk & McDonald, 2004). How might resource concepts play into habitat interactions?

There's a blurring of the divide that seems to assign reason and creativity only to humans. Liane Gabora and Anne Russon (2011), for instance, say there is symbolic behavior in apes, like drumming and signaling gestures. Visual signaling comes through eye and facial movements. Valérie Dufour et al. (2015) go further and say that a chimpanzee can intentionally drum rhythmically with deliberate pacing outside any display context. Encultured consciousness in bonobos forms part of a long discussion by Laurent Dubreuil and Sue Savage-Rumbaugh (2019). Larger brains better organized enable neuronal layering of experiences thus connecting with stronger memory. More than having long-term memory, it has now been demonstrated (Panoz-Brown et al., 2018) that rats replay episodic memory in sequence. If so in rats given similar mammalian evolutionary history, and since an ape's brain neurobiology is close to a human's, then the ape possesses a high degree of sequential mental replay. This means an ape's brain, like a human's, is fully aware of social and cultural contexts. My suggestion, taking this line of thinking further, is how the ape's brain is more responsive than a human's in how ecology contributes to health. Victoria Horner et al. (2010) say that in a way similar to humans, prestige affects cultural learning, where chimpanzees prefer to copy behaviors of a high ranking individual who is successful. Status ranking is classification, an abstract means to sort and identify individuals or types. This, too, apes might apply to certain plentiful or meager habitats.

While the hominid mind, in line with evolutionary psychologists like Leda Cosmides, John Tooby, and Steven Pinker, has inherited modules, Cecilia Heyes (2003) says there were probably other adaptations. As de Waal notes (2001), Heyes is a critic of ape intelligence. Non-cognitive, developmental mechanisms might not have arisen from natural selection. These could be experience-based phylogenetic mutations inherited in a group or ontogenetic genes that get activated in certain environments. Heyes thinks natural selection might be parsimonious in doling out cognitive mechanisms. These various ideas connect because they might explain, in part, our continuity with apes and yet our differences. However, as David Begun (1999) says, homoplasies demonstrate that evolutionary processes are not parsimonious. Gene/culture coevolution is probably the main driver of the difference, where hominin material/art culture helped foster, in an epigenetic as well as genetic manner, human cognitive advances (Tague, 2018). From the perspective of gestalt psychology, behavioral modifications or patterns can be acquired in nongenetic ways. With epigenetics, say Anne Russon and Kim Bard (1996), the advanced cognition of apes draws from more basic forms based on environmental and social factors.

This is not to say apes don't have cognitive domains. Anne Russon and David Begun (2004a) claim that through evolution apes learned to adjust one ability (e.g., foraging) to another (e.g., sociality). For apes, there's a definite correlation between social and ecological spaces. Great ape cognition is spread across a wide plane somewhere between that of other primates and humans. How close to human is ape cognition on that plane? The fact that they can mobilize multiple intelligence modules, like combining tool use, resources, and social strategies (e.g., adult female orangutans feeding together), puts them close to humans. As in human cognitive development, Russon and Begun point out, that of great apes is geared to learning, cooperation and socialization, tool use, theory of mind, abstract communication via gestures, and the ability to mentalize about something in the past or future in another location. Unlike modern humans, apes climb social and cultural steps in forest habitats, not in "civilization." Not all genetic code is necessarily expressed only in chimpanzees or gorillas, regarded as closer to humans than orangutans, say Gisela Kaplan and Lesley Rogers (2000). This is borne out in how orangutans are more insightfully minded when confronted with technical issues at reintroduction centers or zoos. Because genes can be expressed in certain circumstances, the orangutan in its natural habitat does not deal with mechanical gadgets.

INTELLIGENCE AND INSTINCT

My point is that apes evolved higher order mental capacities in order to maintain their forest habitats in a mutually inclusive manner. Intelligence in humans is not unique but, most likely, a combination of super enhancements on preexisting capacities and abilities seen in other species, especially primates and predominantly in great apes. There is no generally accepted definition of animal intelligence, but usually it means how quickly and successfully one solves a variety of problems related to food resourcing, mating, or social behavior, according to Gerhard Roth and Ursula Dicke (2005). Does this mean that intelligence evolved as abilities in special environments and is not comparative across species? Not necessarily, since some psychologists, like Roth and Dicke, believe there is a module of general intelligence along with the concept of learning by association. Mental and behavioral flexibility outside an animal's routine are a measure of intelligence.

Separating intelligence from social behavior is somewhat academically artificial since they can work together. Perhaps the best example of this would be the so-called Machiavellian intelligence or social brain hypothesis where one tries to strategize and out maneuver another socially. Note, though, that Alex DeCasien et al. (2017) question the social brain hypothesis and offer an alternative idea where brain size is determined by diet and not sociality. DeCasien says that like complex social groups the ecological environment can be as demanding in the difficulty of its navigation, the locating of food resources, and the means with which to procure nourishment. Nevertheless, understanding the motives, needs, and desires of someone else in resource competition is a higher order thinking skill which can be used for selfish or cooperative ends. In evolutionary terms humans are closest to the chimpanzee and the bonobo. Going further back in time humans have a close evolutionary relationship to the orangutan. Interestingly, while there are similarities among humans and apes, the social structure of chimpanzees is more competitive than that of orangutans. But this generalization requires elaboration.

For instance, Sebastian Grueneisen et al. (2017) suggest that contrary to previous belief, chimpanzees and bonobos express flexible social-cognitive skills in cooperative, and not only competitive, situations to coordinate decisions. This means that chimpanzees understand and respond to a conspecific's mental state. Social mental processes are neither completely controlled nor automatic but a combination, like the stop/go of grooming, eating, and group movement. Which level of intention is needed for controlled behavior? Some is preconscious and environmentally triggered, like a disgust response. The key factor, says John Bargh (1994), is not simply awareness but how one comprehends that a stimulus could affect future behavioral judgments. One wants to eat but with experience learns how to avoid or overcome repeat competitors. This line of thought in apes might also be applicable to their apparent ecosystem engineering, that is, where ecological interaction is "social." That last thought is difficult for most modern humans to grasp, but consider for a moment hunter-gatherer people and Native Americans already mentioned.

Practical intelligence is related to tool use and how an ape explores its world by manipulating objects. In this regard Kim Bard (1990) studied free-ranging orangutans through the lens of developmental psychologist Jean Piaget. The focus here is on the sensorimotor period in infants, which includes learning cause/effect, imitation, motor intelligence, and object permanence. Some imitation involves taking the mental perspective of another, according to Richard Byrne (1995). This implies intelligence on the level of imitating aspects of the logical structure, impersonating fine details of the sequence, or copying both so that the entire sequence is imitated, as in food resourcing. Byrne says only humans complete the program level (goal and structure) and impersonation (details) in imitation, but this is in terms of, say, an orangutan trying to copy advanced construction work. Apes certainly grasp the program level, and much may depend on the skill imitated relative to the species' evolution and habitat.

Orangutans are capable of imitation by watching. According to Bard (1990), orangutans are ready and competent tool users, for extractive foraging in the wild, but even in captivity. Josep Call (2001) conducted a follow-up study on object permanence in orangutans and found that along with chimpanzees and infant human children all performed correspondingly well. Mostly in Sumatra orangutans use tools to extract food resources in trees (honey and insects) or to break apart nuts. In fact, in prying open a nut, orangutans can first use a thin wedge and then, later, a thicker implement for prying. These tools are not only transported but replaced, and with so many ways to penetrate a food resource, whether in a tree or encased in a nut, there must be a representation in the orangutan's mind in order to know how to proceed and in which sequence. Apes are able to intuit deep forest structures where body and mind are inseparable from the environment. My point is that this monism of mind, body, and environs for apes blurs any artificial distinction between intelligence and instinct. Intelligence is not the same as instinct, of course. For nonhuman primates their evolved nature includes a rational instinct to sustain their native environment since they "know" it intimately coexists with them.

Gisela Kaplan and Lesley Rogers (2000) relate the often-told story that goes like this. In a zoo, give a screwdriver to a chimpanzee, and he will throw it at you; give it to a gorilla, and he will scratch himself; give it to an orangutan, and he will wait until you leave and then use it to escape. This story is based in fact, since orangutans more than other apes will fiddle with mechanisms like locks. Perhaps an ape cousin illustration of this, or not, can be seen in the paper by Jan van Hoof and Bas Lukkenaar (2015). They describe how a captive chimpanzee deliberately chooses and then wields a large stick to strike down a drone, which was being used to take photographs. This could have been an agonistic attack or investigative since once the drone was down the chimpanzee and others poked at it and then handled it for closer inspection. Similarly, Dian Fossey (1983) tells the story of a young mountain gorilla dexterously inspecting her camera and returning it to her. Van Hooff says that in the drone attack, the striking chimpanzee's facial expression was one that indicated a planned maneuver and not a reflexive round of violence. Compare this object use, however, to the even more subtle movements and planning of orangutans who carefully manipulate fabricated metal enclosures to escape.

There are, among chimpanzees, tool sets, sometimes including up to five objects. Chimpanzees who crack nuts have been observed to add wedges to anvils for balance, clearly idea modification and reasoning. Any type of obligate sequence, says William McGrew (2010), to pound, perforate, enlarge, collect, and swab to retrieve honey demonstrates complex cognition. Apes are also competent in using composite tools (the nut hammer on a slab) or compound tools (several leaves compressed to sponge water). Forest materials become an extension of ape mind and body to live in the woods, like early hominins. This is to say nothing about the competency and dexterity shown by great apes in handling branches and other foliage to construct sleeping nests. Young orangutans inspect used nests to learn construction through reverse engineering.

Nesting by apes is exercise of the environment for behavioral adaptation. Barbara Fruth and Gottfried Hohmann (1996) see nesting as the primary bedrock of tool use across great apes, including humans, which evolved into more sophisticated ways to manipulate and so modify the environment. This behavior, says Fruth, goes back to the early Miocene when there were many ape species in forests. Nesting may have arisen from feeding, whether on the ground or in a tree where branch bending or breaking is common to obtain fruits. Orangutans, for instance, build nests in feeding trees and reuse these nests, engaging mental calculation to remain in proximity to food and to ward off smaller primates from eating. Furthermore, says Fruth, nest sites can be seen as hubs where apes communicate about resources. Besides these proximate causes, however, an ultimate cause, says Fruth, is that nest building provided better sleep, health, body homeostasis, and, therefore, cognitive development. Secure rest away from predators allowed for REM sleep (rapid eye movement), necessary for enhanced development of the nervous system and brain health (e.g., memory, cortical development, and learning).

In a study of chimpanzees, orangutans, and capuchin monkeys (Koops et al., 2014), it was determined that tool use/culture derives from "environmental opportunity." This means that social learning stems from ecological factors. Why opportunity and not necessity? Some chimpanzees favor ants where they are abundant and termites not so. Where there are more nut trees, chimpanzees use cracking tools. There is similar behavior in orangutans who, inventing tools, probe tree holes where certain seeds are abundant. Koops says innovation is individual and transmission is social. Caroline Schuppli and Carel van Schaik (2019) count socially learned skills which show a much larger "cultural repertoire" among orangutans, along with ecological correlates for behaviors. Culture is critical to ecologically specific adaptation and likely more widespread than previously thought. As with humans, though on a grander scale, ape ecological engineering, whether through tools (chimpanzees and orangutans) or manual dexterity (gorillas) could have coevolved, as either cause or effect, with intelligence to help manage jungle survival (Navarette et al., 2016).

Across primate species there is no absolute uniformity of intelligence or behaviors. From primate ancestors we apes have inherited a number of mental abilities and social practices, and we've evolved some of our own, in a mix-and-match manner. A hallmark of being human, many people boast, is the mind. But given our close morphological and psychological continuities with great apes, is that only human? According to Darwin (1871, chapter 3) there is only a difference in degree, not kind, between the human and primate mind. Are only humans capable of self-examination? Do only humans consider their actions? Speaking broadly, some philosophers might stress reason over emotion, but emotional response is, too, part of the neurobiology of the mind. Is that only human? Are only humans endowed with free will and somehow exonerated from instincts?

Marc Bekoff (2000) says that animals, especially those more phylogenetically related to us, like primates, probably experience secondary emotions or cortical reflection on how they feel. Researchers like experimental psychologist Sue Savage-Rumbaugh (2000) demonstrate how apes have intentions. Does this mean that only humans experience what we call higher order consciousness? Scientists Todd Feinberg and Jon Mallat (2016) have demonstrated by geological, morphological, and evolutionary evidence that phenomenal consciousness is ancient in origin and has evolved over millions of years from simple organisms. We might understand, then, how basic consciousness would be similar, not distant, in ape species genetically close to our own, and how this consciousness evolved to enable intelligence over instinct. The big difference is in how for modern humans a so-called rational justification for economic and machine progress has superseded any evolved instinct to preserve natural habitats. As readers can see, I'm drawing a fine line between what is human and what is ape. In trying to emphasize the difference, I see apes as having evolved a natural intelligence where their consciousness is geared to forest preservation and not degradation.

According to philosopher William James, is reason an effect of or, according to Adam Smith, a cause of being human? The answer is that there's probably a little of both in a give-and-take ratcheting up interaction. Like other behaviors, what we call reason is evolved and can be seen among other species who plan and deliberate, whether dolphins, corvids, or nonhuman primates. In my estimation, then, ape eco-engineering is both cause and effect of intelligence and social behavior. In a classic biology text, Jerram Brown (1975) carefully delineates the evolution of behavior. Behavioral patterns cross species and are correlated across phylogenetic trees (the sequence of developmental events in time), which demonstrates how evolutionary forces and genetics have shaped many typical actions. Consider how, Brown says, almost all organisms share social behaviors related to agonistic comportment, sexual maneuvers and mating, and kin aid or parental care. What happens is that when a behavior evolves as an adaptation for survival or reproduction to selection pressures in specific environments, that adaptation can affect and have an effect on other behaviors.

Intelligence tests comparing a human child with an ape are human biased and so skew results, especially for those apes with no human acculturation who were entirely institutionalized. For example, an adult ape will not perform well in a gesture test when the ape has never been exposed to an environment where that particular movement is common. In other words, tests for ape intelligence outside of their native, ecological situations, according to David Leavens et al. (2017), account for their supposed shortcomings. One's intelligence is manifested through her evolved cognitive abilities caused by the environs to which she reacts. These abilities can be, certainly, individualistic, but also species based. Cognition, as Arnon Lotem et al. (2017) say, is not physical but emerges from, mostly, learning mechanisms in an environment prescribing adaptations. Cultural innovations, like cooking or dairy farming, modify biological traits, such as jaw muscles and lactose tolerance. In turn, cognitive traits adapt to the new ecological niche. It is human hubris to think we are the only species to have survived and thrived. Most species (probably billions) have gone extinct, and those who remain have sustained their evolution through a combination of selection pressures, whether intelligence or niche construction. A human infant could not survive well in an ape's ecological niche.

Wild orangutans at Tuanan in Borneo demonstrate numerous non-universal innovations where they relate their minds and bodies to the ecology. In fact, there are too many to enumerate here. Some, as cataloged by Carel van Schaik et al. (2006), include sucking ants from hollow tubes, self-scratching with a stick, bunched feeding of ants on leaf nests, nest destruction, smacking sounds while building nests, pillow formation, carrying of leaves, branch cushioning for rest, throat scraping by females to signal infants. These geographically variant behaviors are socially transmitted and, therefore, ecologically cultural. Intelligence is the ability to think outside of the chassis of cognitive mechanisms. As Richard Byrne (1995) points out, innovation is the hallmark of intelligence and the means of cultural change, not exclusively human. Importantly, innovations can affect ecology in a community, broadening the species niche and subsequently leading to speciation, says van Schaik (2006). This work is substantially verified by Meredith Bastian et al. (2012) where innovative behaviors are logged among wild orangutans in Borneo. In this study, two sites were examined where each population was genetically similar but physically separate though in similar habitats. There were patterns of behavior in one site not seen in another, supporting cultural interpretations of orangutan innovation. Some of these innovations include popping sounds and sometimes grinding vocalizations during nest building, nest sharing by a sexually mature male and female, and water drinking from a hollowed base of a rotan stalk. Cultural innovation over instinct gets to the individualism of apes.

Likewise, take, for instance, nonhuman refinements. By the end of the twentieth century it was well established that chimpanzees have culture, forms of which demonstrate a high degree of intelligence. Prominent primatologists in a paper by Andrew Whiten et al. (1999) state with authority that a number of long-term field studies covering 150 years and ranging across Africa clearly indicate there are different chimpanzee cultures because of behavioral variations. One group might employ a technique for the acquisition of food resources another does not, and the cause of this could be behavioral preference. If you are fishing for ants that bite, you'll use a longer stick than usual, so that's an ecological response. If you fish for the same type of ants in your group using a twig rather than a long, slender leaf, that's cultural and probably learned by copying. These are not Whiten's examples. In his study, Whiten says there are thirty-nine varied behavioral patterns covering, for example, tool use, grooming, and courtship. If there's no ecological explanation for the behavior, then it is cultural. Note, however, that the ape "culture" is geared to forest sustainability where instinct and intelligence for deep-rooted preservation meet. Nevertheless, even many ape cultural advances are rooted in some form of food resourcing and are, therefore, tied to eco-engineering. More recently, Caroline Schuppli and Carel van Schaik (2019) broaden extensively the notion of an ape culture that is not based on geographical exclusion and aver that great apes are as cultural as humans. Aspects of evolution can occur culturally, as many biologists and evolutionary psychologists, such as Eva Jablonka and Marion Lamb (2005), Alex Mesoudi (2011), and Mark Pagel (2012), have written. The preceding sections bring us to this question.

CAN APES THINK?

The answer is affirmative, though others, noted below, differ. Anthony Quinton (1967) shows how philosopher Karl Popper tried to discriminate the empiricism of science from metaphysics and myth. Why is this relevant? Popper distinguishes between the psychology of knowledge (empirical facts) and the logic of knowledge (relationships). Apes are capable of induction which tests hypotheses (e.g., is that limb strong yet flexible?). Induction is synthetic. Consider, for example, trial and error. Moreover, apes are competent with deduction which explicates hypotheses to testable consequences (e.g., will that flexible limb support my weight to the nearby fruit tree?). Deduction is analytic; one plans from observation. Where induction quantifies facts, deduction puts facts in order. Great apes deduce, particularly orangutans with their sharp attention to mechanical detail, in their ability to put together facts and then execute a plan. This has been evident time and again in zoos, since orangutans are known for their ability to escape from enclosures by taking advantage of accidental or premeditated access to tools.

What do apes think about and what for? Barbara Harrisson (1962) relates several Houdini-like maneuvers. Harrisson's young orangutan charge, Bob, always toyed with the panels of his wooden enclosure, managing to escape. Later, when he was transferred to the San Diego Zoo, he did the same with his wire cage, taking his blanket and toys with him as if he had no plans of returning. Once the cage was repaired and Bob reinstated, he began to plot a new escape the next day, and was successful, ending up with the penguins. Even when Bob was then placed in a heavily barred enclosure, previously used for grizzly bears, he managed to manipulate or twist the bars enough to escape, found early the next day in a feeding room where he had partaken of many delights during the night. This young individual demonstrates not only the ingenuity of orangutans to engineer an environment but also their conscious intention to be at liberty as free agents.

Michael Tomasello (2014) says human thinking, which derives from all expansive and ever increasing cognitive culture, is unique because it is cooperative. But how prevalent and regular is human cooperation? More so, cooperation for what ends? Would one say a team of people clubbing baby seals to death is "cooperative"? Tomasello sees culture from a human perspective and as social coordination. No one would disagree that human cultures are more advanced and widespread than those of great apes, but that was not always the case. Why is it that the basic toolkit of early hominins, from *Homo habilis* up to *Homo erectus*, was somewhat similar across regions without major change? Yet those species are direct descendants of ancient human ancestors. There is archaeological evidence, say Julio Mercader et al. (2007), that chimpanzees used stone tools over 4kya for food resource gathering before the advent of human agriculture in parts of Africa. This is significant because it suggests that percussive tool use derives from a common ancestor and is not a hominin invention. Furthermore, wild male orangutans exhibit classic signs of cognitive planning, self-control, and mental time travel by announcing the next day's travel route through a long call the night before, thus attracting females and repelling low ranking males (van Schaik, 2013). There's also the case of the wild chimpanzee who ate part of a tortoise on Monday, put what remained in the crook of a tree, and then came back on Tuesday to finish his meal (Pika et al., 2019).

Displaced reference, evident in, for example, the long call, is an evolutionary linguistic ability where a communicator points to past, future, or distant events or things. Similarly, Adriano Lameira and Josep Call (2018) say wild female, Sumatran orangutans stifle alarm calls in the presence of infants for up to twenty minutes until the danger of a predator has passed. I see this as evidence not only of a mother concealing the whereabouts of herself and offspring but as a way to instruct a youngster in self-control. The delay is explainable since infants are highly susceptible to big cats that lie in wait and, so, might not be visible after a first appearance. In addition to deferring the alarm call, the mother with young offspring uses the opportunity to teach the infant about predators. Previously, it was not thought a nonhuman animal could engage in a time/space displaced response, say these authors. Their research demonstrates how communication delay, for a speaker and listener, involves higher cognitive functions like intention, memory, and theory of mind, elements of a morally individual person.

What makes me a little uncomfortable is when respected thinkers label all of human abilities as unique. Surely we are special, they say. Eileen Crist (2019) fittingly states that some Western, industrialized human societies have promoted their own "uniqueness" by indoctrinating and conditioning others. Evolving adaptations is not unique to one species, nor is there any special essence to any species. Selection pressures work in different ways on different genes in different ecosystems. Many species have survived and thrived longer than humans, like the horseshoe crab which has been around, granted with little change, for about 400my. Were humans to annihilate themselves and many other plant and animal species, some, perhaps the horseshoe crab, would endure. As an experienced researcher Tomasello (2014) agrees that great apes understand the intentions of another and can problem solve, but they don't do so as a collective. There's a degree of mental coordination in humans not quite seen in great apes. While this is true, we also need to take into account the ape land ethic as moral individuals to sustainably eco-engineer their domains. In their environment of evolutionary adaptation that includes spaced feeding and infrequent births, cooperation on a grand scale obviously counted less for apes. Consider which species was content in its evolutionary biome and which has already "cooperatively" endangered the existence of the entire biosphere. Who truly evinces a land ethic? Besides, apes do intentionally cooperate. Barbara Harrisson (1962) tells the story of how ex-captive, young male orangutans Nigel and Frank help each other build a single nest together.

The ability to massively coordinate thinking has some serious downsides and almost seems outside the human scale. Surprisingly, Tomasello (2014) insists that only humans can think objectively, engage in social reflection, and evaluate thoughts relative to social norms. But Frans de Waal's early book *Chimpanzee Politics* (1982) shows that apes, too, are capable of controlling their actions in some evaluative and representational manner. This is no small matter and hinges on the larger picture of moral individualism. Philosopher J. L. Mackie (in Rachels, 1998, p. 210) says thought is sensitivity to the world, so mental representations have moral implications. Where Tomasello says chimpanzees are competitive, Wolfgang Köhler (1925) says they are, following Mackie's phenomenological thesis, frequently engaged in watching each other at environmental tasks. Granted apes can be competitive in their individuality, but that observation should not be an indictment against their intelligence since some routine human aggression (whether verbal or physical) might be a hominin adaptation and not inherited from a common ancestor. As evolutionary psychologist David Buss (2008) points out, there are really only two species where males initiate territorial aggression: humans and chimpanzees. Some great apes, unless provoked, are quite serene, like the orangutan, or gentle, like the gorilla, or peaceful, like the bonobo who is very close to humans genetically.

Köhler (1925) does not look for defined ape/human equivalencies with intelligence. More sensibly he tunes into approximations of intelligence traits. This is because, as Köhler notes, intelligence comes from imitating the *substance* and not merely the *form* of another's actions. Köhler, in his wide observations of chimpanzees working on problem solving tasks, says there's a correlation between intelligence and dexterity. Such mind-body coordination unquestionably relates to ecosystem engineering. Some of the more clumsy chimpanzees were not so smart. Like humans, while chimpanzees immerse themselves in trial and error, in some cases they clearly think prior to acting.

Dorothy Cheney and Robert Seyfarth (1990) follow Köhler (1925) and say primates learn not simply by behavioral stimulus response but by insight. Also, they agree with Konrad Lorenz who emphasizes how mental perception contributes to adaptation. Psychological representations, particularly for primates, deal with conjuring the mental states of others. That's standard, but because great apes also spend stretches of time alone or in small, diffuse groups immersed in green spaces teeming with a myriad of life forms, mental representations are strongly eco-centered. Indeed, in larger primates visual acuity supersedes olfactory sensation, no doubt to enhance accurate observation, evaluation, and memory not only about another's behavior but also regarding food sourcing in space and time. Apes don't just know *how*; they sense causally that one behaves so or can behave so because of rank, status, desires, or needs. This mental activity of understanding causation is used to engineer a habitat as well. Some might say this is not cognitive behavior but simply a way to make associations.

At any rate, if an ape can step outside of her *self* to grasp the thinking of another, that's intelligent behavior with a social goal. Such multiple dimensions would also have benefits in terms of placing oneself in the physical environment. Chimpanzees and orangutans have the cognitive ability to classify objects, like tools, according to their function, so they probably make similar hypotheses about the behavior of others by mental inference. No doubt they also index trees, plants, insects, and objects for immediate or future use. As Cheney and Seyfarth (1990) suggest, compared with monkeys who possess mental domain specificity and are driven only by ecological context, great apes have general purpose intelligence and are reflective about their own knowledge, especially in terms of social situations but not without accounting for the contents of or movements in their forest homes.

Frans de Waal (2016b) rightly asks if we are even wise enough to gauge animal intelligence. There's a false dichotomy between human/animal. Rather, intimates de Waal, whatever we label as human intelligence is simply one point on an extended evolutionary plane of cognitive abilities across species. We see such differences in the application of intelligence among great ape species depending on their group size, if there even is a group, and how resources are obtained and shared. Group implies aggregate members, but typically wild chimpanzees fissure into small units during the day, though they are sensitive to the whereabouts of others, and then fuse at night. Primate intelligence evolved because of the mental demands of multidimensional social life in a threatening but yet abundantly rich forest environment. There are precedents of rational consciousness in other species, especially in great apes, evident from their cognitive behaviors.

The difference is that the environment of evolutionary adaptation adjusted cognition for a variety of species in different habitat scenarios. Because apes are large and social, their thinking concerns other individuals relative to environmental resources. How any organism processes information is cognition. Orangutans are deliberate where bonobos and chimpanzees can be excitable. That's a generalization, though, for as de Waal (1982) relates, chimpanzees can often reflect before they act. They think with purpose and can manufacture spontaneous, creative solutions based on reasoning from experience. Whereas information processing is cognition, output and outcomes are manifestations of intelligence. Contemplation points to how great apes have mental representations in mind. Why wouldn't apes construct mental thoughts, considering their daily environmental challenges and social negotiations? My point is that apes employ insight and understanding across a range of plant species, not just with each other.

As some might say mind is paramount in humans, others say language is the quintessential human factor. But cognition is not dependent on language. Arm and hand gestures are exclusive to humans and apes, not monkeys. Wernicke's brain area, relevant to speech development, exists in apes as well as humans. The FOXP2 gene, related to speech development and language, exists in humans and birds. Regarding the intelligence of birds, especially corvids like crows, they are generally known to make and use tools, engage in rituals over their dead, use cars in traffic to crack nuts, recognize faces, and offer gifts to humans. A magpie, in fact, can pass the mirror self-recognition test. Birds don't possess mammalian cortical structures but yet have intricate networking that might replicate a cortex. These common facts simply demonstrate how some human abilities, like language, are not just uniquely pronounced only in *Homo sapiens* but derive from ancient precedents, primitive design patterns, and extant species. Conventional wisdom dictates that without language there is no cognition, but that is patently false. Chimpanzees and elephants have excellent memories, for instance, and memory is part of the cognitive process.

Natural selection is responsible for cognition, but cognition is not the product of natural selection since the ultimate outcomes of cognition are what count in selection, like how to use memories. Tomasello (2014) says that apes, especially chimpanzees, bonobos, and orangutans, are capable of inferential thinking or imagining what could be one of several results in a situation. They monitor their own thoughts and calculate future rewards to the point of patiently holding back for anticipated gain. This ability is environmental as much as social. Mental monitoring is advanced enough so that, Tomasello says, apes understand when they don't quite know enough regarding a situation or event and need to wait while considering alternatives, including inaction.

Instrumental rationality (e.g., cost effectiveness evident in many species) in achieving practical ends probably dates back to *Australopithecus* or a shared ancestor to hominins and chimpanzees. Yet the major difference between apes and humans, says Tomasello (2014), is that whereas ape cognition is for competition, human cognition is for cooperation. Does that line of thinking imply that apes are without a communal value system and only self-interested? Tomasello simplifies complex behaviors and ignores ape intelligence that arose because of cost/benefit responses to environmental constraints or stimuli. Take, for instance, how Biruté Galdikas (1989) cogently makes the case for orangutan tool use through their cognitive ability related to locomotion. Trees of different thickness and heights, mother trees and saplings, supporting branches, and vegetative flexibility are used judiciously like tools to manipulate movement. Apes figuratively cooperate with the "mind" and body of the tropical forest. Orangutans feed at the top of the trees which get sunlight and bear the best fruit, but these forest people don't jump and are aware that some branch extensions and lianas will not support their weight. Such considerations carry mental evaluations and turn their arboreal movements into a physics class. Though technically neither cooperative nor competitive, this intelligence should be seen as taking stock of fortuitous physical circumstances for self and perhaps a dependent offspring. Besides, if the orangutan successfully evolved to be semi-solitary, why fault the species, according to Tomasello's line of thinking, for not being "cooperative"?

As reported by Gisela Kaplan and Lesley Rogers (2000), weight movement in trees demonstrates a high intelligence related to problem solving. Climbing by such large creatures is an evolutionary adaptation (feeding, predator and parasite avoidance) that affected brain growth. Concentration on how one moves her body in open space using all four limbs while carrying a baby indicates a very high degree of self-awareness. Contrary to how researchers like Tomasello see apes as competitive, here is a routine showing how female orangutans with young offspring will at times feed together. In fact, I'd further argue that orangutans "cooperate" with the challenges and opportunities of their environs. It's difficult to determine if Tomasello's (2014) goal is to make a case for what he terms human uniqueness. No matter, since even he admits how human evolutionary history is very close to that of the great apes. In contrast to apes, there is much more cooperative behavior among modern human foragers. It could be that when hominins first appeared they faced stiff competition for food resources and, therefore, evolved adaptations for shared, cooperative intentions that enabled them to monitor the thoughts of another.

Having said that simply demonstrates how apes and hominins evolved differently to fill dissimilar ecological niches in their own environments of evolutionary adaptation. As stated by Juan Carlos Gómez (1990), a gorilla knows a human as an agent and knows the human can similarly perceive a problem where both can collaborate, demonstrating intentional communication. What else do they sense and what can they figure out? Leif Cocks

(2016) tells the story of how Hsing Hsing intuited intention from accident. This orangutan, after careful observation of keepers, not only discerned how to escape from his enclosure but was intent on letting the other orangutans out, sensing that they, too, probably felt confined. He was stopped before he could complete his compassionate mission. Anyway, because Tomasello (2014) does not find other-person thinking in apes, he does not believe they are normative creatures. Importantly, as Jonathan Dancy (in Rachels, 1998, p. 228) would say, moral properties need minds. Apes clearly do consider another's behavior, needs, and motives and will act on them. Granted that action could be self-directed, but this only means their social *mores* differ from the overtly cooperative norms of most humans and some other species. Let's say "overtly" because, contrary to the altruism suggested by human cooperation, many people often act deceptively/cooperatively out of self-interest for themselves, kin, or group. Cooperation, then, can be competitive. Altruism can be competitive. Indeed, competition can very well engender social behavior, even among humans.

Generally speaking, primate competitiveness for resources, a niche, or reproductive opportunities has shaped their temperaments. Primates can be manipulative to satisfy their needs and desires, from intimidation displays, wild tantrums, or peaceful grooming. Primate cognition evolved in an environment where one tried to outcompete another for status or a variety of resources. Saying this is important because, as Brian Hare (2001) points out, the difficulty of effectively evaluating an ape's behavior, especially a chimpanzee's, is compromised since testing is done in a cooperative environment. For instance, in competitive situations for food chimpanzees are aware of what others know. The bottom line is that apes indeed think, and such cognitive functions need not always be determined by instinct-driven competition.

Let me take this analysis to the next level and consider how apes form a theory of another's mind. In turn, I will relate that mind reading capacity to how they interact with their forest habitats.

THEORY OF MIND

Advanced ape mentality has been studied systematically since 1978, and the debate about great ape theory of mind has been active for quite some time. While executive functions come into play with theory of mind, so does the gut (literally and figuratively) since it's a guessing game often about resources. Theory of mind is a psychological means where one attempts to enter into the mind of another. Essentially, theory of mind is a simulation where one tries to read another's mind based on her own experience. The guessing involves detecting another's potential beliefs and actions for strategic, cooperative, competitive, and even manipulative means. Theory of mind is joint attention and can attempt to anticipate *what if*. Humans comprehend another's mental state via language; but people also follow another's gaze, read hand gestures, bodily movements, and facial expressions. So when we attempt to read another's mind we make a theory about his thoughts, feelings, beliefs, or intentions. Great apes do this too.

Readers might justifiably ask what have these sections on ape thinking and theory of mind have to do with ecosystem engineering. Emotions and intelligence get to the heart of moral feeling and ethical individualism which define someone. A person willingly acts in a community of kin and others; ergo, the realization and fulfillment of a forest person's evolved emotional and intelligent social instincts depend on the viability of her physical environment as well. In distinction to most modern humans (in towns or cities composed of constructions that separate people from nature), an ape's gestalt is naturally connected and organically contributes to her *Umwelt*. There is no psychological division between self, habitat, and other. The mind of an ape is part of other ape minds in a tropical forest. I'd go as far as saying that, at times, an ape tries to read the resonating mind of the forest. That is, ape canonical neurons respond to and define objects in a type of motor cognition, offering one choices about what to do. Canonical neurons involve how a primate processes what she sees and then computes how to use it, like a tree limb in different situations. I'm not sure robots or computers can yet boast the full-bodied psychology of forest apes or produce a land ethic.

Having a theory of mind is critical to any ethical status of personhood, says Richard Byrne (1995). Frans de Waal (2016b) does not like the "theory" component and asserts, rather, that any guessing is bodily and not abstract. While abstraction is certainly involved, for the purposes of my discussion it should be evident that embodied cognition is paramount. De Waal prefers to say perspective taking, as in empathy and body reading. De Waal provides illustrations of what he means and shows the extent of a chimpanzee's repertoire. For example, one can disguise what he knows about a food location. Or one can read another's body language to decipher information about a nutrition source. There are also examples of how chimpanzees will observe caretakers in order to obtain aid or food. Nonetheless, de Waal (2016b) says that regarding theory of mind, chimpanzees read other chimps in their own environment, not in a research lab.

Great apes consider another individual's knowledge and understand the difference between an accidental error and intentional harm. Byrne (1995) goes on to say, in answering the question about the usefulness of theory of mind, that one is able to grasp that others have knowledge and desires different from one's own, implying moral considerations. This means much in predicting and reacting to another's anticipated behavior and cooperation.

The elements of differentiating between good and bad behavior are implied here, since using any theory of mind can involve deception and manipulation, whether self-serving or for the group. Theory of mind is, therefore, essential to moral individualism.

Important here is that if an ape knows a conspecific also has thoughts and beliefs, then there is a recognition of self and other identities, critical for a social emotion like empathy. Beyond the philosophical notion of "the other," apes are aware of the forest matrix of other entities, plant and animal. There is some level of control in applying theory of mind. While we eagerly guess at another's feelings we can regulate the degree to which it is active. There might be a dual action process in theory of mind: forming a representation of another's thoughts and predicting another's behavior based on that representation. Humans often guess incorrectly as they project their own thoughts and beliefs onto another. So do apes and why, in theory of mind tests, there is never 100 percent accuracy. People make assumptions. So do apes, and such mental interactions over mere stimulus response are adaptive and probably, in part, explain the evolution of consciousness. Clearly apes are more conscious of their ecosystem habitats than most modern humans are of "nature." An ape's mind is connected to the forest.

No exhaustive history of ape theory of mind research is given here, but I will cover some recent, pertinent highlights. For a more comprehensive survey, see Kristin Andrews (2016). There were some early scientists who studied apes and took seriously their mentality, like Robert Yerkes (1916), who found apes to be mechanically minded, and Wolfgang Köhler (1925), who designated them as insightful. By 1978 David Premack and Guy Woodruff were able to ask candidly whether or not a chimpanzee has a theory of mind. To answer the question these researchers showed a fourteen-year-old adult female chimpanzee from Africa, Sarah, thirty-second video tapes of a human actor trying to solve certain problems, like reaching for food, escaping from a cage, trying to generate heat, and attempting to fix a phonograph. Along with each of the video tapes the chimpanzee had some photographs, one of which contained an image of a solution to the problem (e.g., a key for a locked cage).

Consistently, Sarah chose the correct photo to solve the problem. This means that the chimpanzee inferred that the human in the video had intention and belief. The researchers concluded that Sarah solved the problems (twenty-one of twenty-four tries with eleven of twelve more complex problems) by using theory of mind to understand what the actor believed or desired. She even exhibited empathy in offering aid to the actor. In further tests the choices of the chimpanzee were affected by how well she knew or liked the human actor. All of these processes and outcomes point to a richly conscious mental state. Premack and Woodruff show, then, that theory of mind is not necessarily taught but learned naturally, though suppression of inferences are learned. Apes are especially keen about guessing motives, needs, and desires. Is it hard to comprehend, then, that for a forest person such mental activity has bearing on how she imagines and interacts with all of the vibrating life in her ecosystem habitat.

Much later Daniel Povinelli et al. (1990) tested chimpanzees to conclude that the apes are more than capable of modeling another's point of view. Such perspective taking enables manipulation and deception. De Waal (1982) tells the story of Dandy who, while others were snoozing in siesta, completed the execution of his plan to retrieve hidden grapefruits for himself. Povinelli further says that apes are likely capable of differentiating between one who knows, by seeing what happened, and another who does not know. This means that apes can understand different states of mind for others of the group. They try to apprehend other minds. It should be noted that initially Povinelli said there probably is some great ape theory of mind; but later he came to believe full theory of mind is only human. Compare with Michael Tomasello, who first said there was no ape theory of mind but now sees some of it in great apes.

José Luis Bermúdez (2003, p. 9) sees behavioral "input-output links" or stimulus responses that don't depend on intermediary representation, unless recalling one that is stored. All species of infant apes undergo a long process of schooling so that they can create and store not only mental representations but appropriate behavioral responses. This "thinking without words" for Bermúdez does not require any explanation. I don't believe, however, this is the same as Cecilia Heyes (1998) finding no convincing evidence of theory of mind in primates. She does not believe that apes know anything about the content of their mental representations or think about the mental states of others. How, then, could they have survived in ecologically complex social groups? On the one hand, Bermúdez confirms my notion of ape ecosystem engineering, where they guess at and respond to objects in their ecosystems, while Heyes, on the other hand, might discount such neuronal activity as mere instinct since she's questioning any human theory of mind ability in apes.

There is plentiful evidence, though, to overturn Heyes' doubts. For instance, reared from nine months, and after a few years, male orangutan Chantek developed a vocabulary of 140 signs as well as creating new signs of his own, says H. Lyn White Miles (1990). Like a human child learning and developing language, this orangutan person was able to cultivate referential skills because of his communicative and cognitive abilities. Importantly, Chantek comprehended the abstract nature of signs, representations of reality, and hence evidence of mind. Chantek was capable of stringing together stones or beads into jewelry which he'd then gift to someone. This ape behavior is not surprising to some, as seen in philosopher Mark Rowlands (2019) who mounts a convincing argument for animal minds. Additionally, Leif Cocks (2016) tells the story of the male orangutan Hsing Hsing whom he tested and treated for diabetes by finger pricks and shoulder injection. The orangutan person was more attentive to and learned readily from Cocks' body language and movements. Even though chimpanzees and orangutans can pass a self-recognition test, Heyes (1998) is skeptical of those results and other indicators of ape mentality. Imitation is overrated, she says, because role taking might simply be familiarity. Deception, she believes, could occur by chance observation of cause and effect and not because of mental powers. Heyes says, finally, that perspective taking, creating a relation between seeing and knowing, is questionable in apes.

Surely self-awareness is implicated in theory of mind, and would make sense for the survival of a large forest creature interacting with conspecifics for resources. Contrary to Gordon Gallup (1970), György Gergely (1994) says that cognitive capacities for mirror self-recognition are not necessarily connected to theory of mind. For Gergely, theory of mind consists of:

1. Concept formation and inference;
2. Grasping variables of any perceived reality;
3. Making hypotheticals via reality;
4. Forming meta-representation or understanding how to mentally display any reference to a sensation and the difference between the two;
5. Comprehending the relationship between a mental state and action.

Although Gergely's paper appears in a book dealing with self-awareness of humans and nonhuman animals, he does not say whether or not his template applies to great apes. Let's assume it does, based on relevant evidence accumulated here. Furthermore, these elements relate to how apes, as forest eco-engineers, apply a theory of mind to the actual, physical environs, consistently moving, watching, and guessing about objects and sounds.

Daniel Povinelli and Jennifer Vonk (2003) ask if an ability to conceive a mental world is restricted to humans. Do only humans think about their mental states? Their answer is that apes abstractly represent another's behavior in order to classify others and other behaviors so as to plan and predict future outcomes. These mental representations can also be applied to one's own behavior. In this article, however, Povinelli and Vonk are skeptical about ape theory of mind and seem to believe it is uniquely human, as if forming mental abstractions about another's behavior and theory of mind are separate. This could be true, but based on Christopher Krupenye (2016) and David Buttlemann (2017), to be discussed soon, one is inclined to say no. Certainly human theory of mind might be of an advanced degree on the same plane, likely because of a coevolution of material/art culture and mental representations

(Tague, 2018). Some skeptics might say apes don't really deceive or manipulate another's behavior and only comprehend a stimulus response, but that minimizes their high capacity for socialization.

The problem with the line of thinking that negates advanced ape mentality, even given the papers discussed so far, is that it favors behaviorism. Great apes are not just copying machines but have innate mental modules evolved because of adaptations to complex social and especially, as I have been arguing, ecological environments. Humans don't simply link behaviors together like keys on a chain, nor do apes. People think about those behaviors using very similar mental representations and theory of mind, just as apes, based on, among other things, their environment of evolutionary adaptation and their current physical surroundings in terms of embodied cognition.

Are there degrees of theory of mind? Possibly, as consciousness itself is an evolved behavioral mechanism and varies in intensities (Feinberg & Mallet, 2016). Even if theory of mind is graded, apes have it since theirs is on such a close evolutionary track to that of humans. This logic, of course, could lead to a type of infinite regress, but that makes sense since adaptations evolve from simple to complex (e.g., a light conductor which becomes an eye or a swim bladder which becomes a lung). Benjamin Purzycki (2005), however, says great apes don't have human theory of mind but something in line with their own evolution. Coalition forming in chimpanzees suggests basic theory of mind. Helping behaviors in apes also suggests theory of mind since they indicate a sharing of intentions. Bees manifest helping behaviors, too, and they evince simple perspective taking when one signals others about resources. Relatedly, apes can hold mental representations, evident in how after a confrontation a chimpanzee will wait for a chance at reconciliation. We also know, especially with chimpanzees, that they will deceive each other to hide or obtain prized food. Purzycki is not an ape theory of mind skeptic like Povinelli, Vonk, and Heyes; instead, he sees theory of mind along a spectrum, where human access to and mental representations about another's motives, needs, and desires are more advanced.

Thirty years after the Premack and Woodruff (1978) paper Josep Call and Michael Tomasello (2008) offer an equivocal answer to the question of great ape, especially chimpanzee, theory of mind. Acknowledging that researchers have clearly demonstrated an ape's understanding of another's intentions and distinct identity, there is, at the time of their writing they claim, no evidence that an ape comprehends false beliefs. That is, an ape, they say, cannot understand that another holds a belief different than his own. They see apes as perception-goal automatons. This attitude is too simplistic given, even among the semi-solitary orangutan, the politics around status or hierarchy, kin relations, and shared ecology. Apes think psychologically about the meaning of each other's movements. Chimpanzees, because they are highly mobile and free-ranging, are not often together and, in some cases, days could pass before meetings of community members (Goodall, 1986). In this case, an understanding of another's presence/absence in the forest ecology, as if the environs extend the body, is paramount. Undoubtedly ape theory of mind is linked to their movements among plants and trees, as a group, a dyad, or even when alone.

Of course great apes have a psychology. They behave in ways that demonstrate they are continually thinking about how others might perform. In the forest, these speculations probably extend to other organisms. Chimpanzees interact often and so have contractual obligations. This is not to say that the semi-solitary orangutan is not social; she is just more independently minded than humans and chimpanzees. Overall, wild, adolescent, female orangutans are actually quite social, according to Biruté Galdikas (1995a), and subadult males will follow or associate with these females, probably for reproductive benefits. Granted, in their natural habitat, like the environment of their evolutionary adaptation, they will behave differently than they would in a zoo or lab. In a review of numerous studies, Call and Tomasello (2008) admit that chimpanzees could possibly grasp another's intentions. Beyond mere surface level behavior, apes attend to and understand the psychological behaviors of others. Nonetheless, Call and Tomasello insist that chimpanzees are incapable of estimating another's actions from what the chimpanzee knows and the actor doesn't. They offer a narrow theory of mind to chimpanzees. We now know this conclusion is not wholly accurate. Great apes grasp that others hold beliefs and mental representations different from their own, as Krupenye (2016) and Buttlemann (2017) show.

Deeper skepticism about ape theory of mind can be found in Alia Martin and Laurie Santos (2016). In an otherwise thorough paper, these authors admit that primates, broadly speaking, can track the awareness of others but, without an ability to detect false belief, do not possess true theory of mind. False belief on the simplest level means that one knows his mental representations might clash with reality. How could one even suggest that a forest-living ape does not possess that capacity? Of course apes conjecture about the ecosystem plants and animals for that is how they have evolved survival and reproductive adaptations. The problem is that some researchers completely take the primates out of the forest.

The human attitude should not, however, be to measure apes to human criteria. People don't measure their children's inabilities to climb tall trees or lack of physical strength against those in apes. Martin and Santos (2016) say there is no evidence that apes decouple awareness from representation to form a *what if* situation. Based on the escape stories of orangutans outlined by Leif Cocks (2016) and arboreal clambering described by Povinelli and Cant (1995), this is not accurate. The example of decoupling reality from representation used by Martin and Santos is the false belief test, knowing others believe something dissimilar to one's own belief, which they say apes cannot accomplish. They admit, anyway, that apes can infer.

All of which brings me to how, according to Christopher Krupenye et al. (2016), great apes indeed employ advanced theory of mind, which means that they understand another agent can have a false belief. This innovative study included seven orangutans, fourteen chimpanzees, and nine bonobos. The authors acknowledge that in previous tests to evaluate ape false belief ability, in spite of their high intelligence, there has been failure by the apes and researchers. Perhaps errors were made as humans often mistakenly guess another's intentions. Or previous tests by humans with humans on apes discounted the disconnection from the apes' natural habitat and environment of evolutionary adaptation. So the Krupenye trial made use of an infrared camera device to measure anticipatory looking. The results show that at least these three species of great apes, like humans, anticipate an actor's response to his own false belief by their looking in a certain direction where the actor falsely believes an object to be.

For example, simplifying the Krupenye narrative, two actors are in a room with both a black and white bureau, and actor one sees actor two place a book in the white bureau's drawer. Apes witness all the action. Actor one is then escorted out of the room while actor two moves the book to the black bureau's drawer. Returning to the room actor one will look where he believes the book has been (where he last saw it) and great apes' eye movements have been tracked by Krupenye to demonstrate that they comprehend how another can have a false belief. The apes know that actor one will look in the white bureau's drawer even though they know the book is not there. This shows that apes, like humans, can attribute mental states including intentions, desires, knowledge, and beliefs to oneself and others.

The implications here are immense, since it means that apes are not guided only by awareness of others in real time but also by beliefs about how others can act and the knowledge that beliefs of actors can be false. Moral textures can be implicated in mind reading. Based on their physical and psychological experiences, apes will guess at the mental states of others. They read minds as they reinterpret their own sensations through another, and if more descriptive than predictive there is, nonetheless, probative value. If this is not moral individualism with autonomy and responsibility, that is, personhood, then what is?

As a side note, let's ask if episodic memory indicates theory of mind? Michael Mendl and Elizabeth Paul (2008) might say no; but they also indicate that the absence of perspective taking or mental time travel does not diminish an animal's emotions or ability to have affective states with the environment. Moods and emotions connect with perceptual variations, and Andrew Crump et al. (2018) seem to say specific stimuli can be determinant, which seems to bolster the use of eye tracking. My point, of course, is that given apes' highly advanced mental states (and not just lumping them in with "animals") renders them close to humans, with the difference that their affective states, perspective taking, and mental backward-and-forward thinking are, when not socially engaged, tied to their ecosystem engineering in a challenging forest network.

In humans, full-fledged grasp of false belief occurs after the age of four. A skeptic like Laurie Santos, quoted by Morell (2016), says that there are already too many papers that negate apes passing a false belief test and that Krupenye merely confirms that apes can predict behavior. People would not put such emphasis on tests with human children and often indicate how some children are not good at taking tests concocted by adults. In a supportive opinion piece, Frans de Waal (2016a) lauds the findings of Krupenye's study. However one interprets Krupenye, it is significant that apes know when others demonstrate a perspective not in line with reality. Are apes, then, really able to comprehend another's mind? In a speculative but not wholly illogical manner, following the course of my argument, I ask if apes can grasp the mind of the forest? Why do people doubt this and yet erroneously accept artificial intelligence in computers which are incapable of guessing my intentions or beliefs? Catherine Caruso (2016) reports that researcher Robert Lurz (2012) says apes are unable to tap into another's thoughts. He says that Krupenye only shows typical ape behavior. How does one know what's typical? In a lab, a zoo? Based on a study of academic papers and field narratives, how could one not agree with the Krupenye findings?

Social cognition is vital to interpersonal and group relationships, all of which carry implications for moral individualism. Since Krupenye there have been skeptics, of course, notably Cecilia Heyes (2016) who says in a writing authored only by her, that Krupenye does not prove mentalizing but only submentalizing or predicting behavior. Heyes means that the apes focused on what they saw and not on what they believed. Heyes speculates that the eye movements so carefully tracked among thirty apes over many trials in the Krupenye research was simply generated by a memory sequence of what was seen in context. It is worth noting that Krupenye's study includes prominent researchers like Josep Call and Michael Tomasello who teamed up with other authors on the earlier ape theory of mind papers. Heyes claims that the apes did not focus on what they believed but only on what they saw indicating low-level psychological processes and not the higher consciousness called theory of mind.

Heyes was then challenged by Fumihiro Kano et al. (2017) who say that predictive mentalizing is not an adequate explanation for an ape's belief-based anticipation of the actor's movements. Kano did a test following Heyes' advice, by not using actors but inanimate objects. Sure enough, the apes were far less interested in the location of objects than the movement of an actor based on their belief of his understanding of the placement of an object. This result is not surprising since primates, specifically, and great apes, especially, are mentally motivated to guess the thoughts of others. Like humans, apes know that behavior is not necessarily driven by true knowledge but by beliefs about reality, sometimes false beliefs. Such cognition is more than submentalizing and might even affect how they interact with the uncertainties of a tropical forest itself. Further findings by Kano (2019) show how great apes, without behavioral cues, use past, perceptual experience to anticipate an agent's action. This might attest to mentalizing extending beyond an ape's perimeter, meaning there could be implications for eco-engineering when they see the forest and its parts as interacting "agents."

Employing an interactive helping task, David Buttelmann et al. (2017) confirm the Krupenye findings. Great apes can differentiate true and false beliefs when they assist another. Here, too, we are not surprised given the long evolutionary history of the advanced social nature of apes. Still, the question comes up about whether Krupenye's apes predicted well or truly grasped the actor's false belief. Buttlemann's study is not based on looking but is behavioral, a key difference. Why? If an actor is trying to accomplish something while holding a false understanding and is successfully helped by a subject who has true understanding of the situation, then the subject comprehends another's false belief. This is not surprising given, broadly speaking, the helping nature of great apes.

The key question, according to Thomas Bugnyar (2017), is whether an ape or other animals, including Rhesus macaques or birds like scrub jays and ravens, are able to represent the belief of another and so know another's actions can be guided by a false belief. Otherwise it's fairly well accepted that apes possess aspects of theory of mind. Bugnyar says Krupenye is statistically reliable based on the original number of apes (forty) where twenty-nine participated, with little results variation between species. He says studies need to make sure there is no submentalizing present where from experience one knows an agent will look for an item in its last location, as presented by Kano (2017) against Heyes. Bugnyar's ideas, however, are in contrast to some who might say that while apes could be able to represent a belief, they still might not understand what they "know," but they sense something.

At least one cognitive philosopher (Andrews, 2018) has raised the question about whether or not apes understand the content of another's beliefs. Here are the constituents of a belief:

1. Proposition, or the meaning
2. Referential representation
3. Action, based on desire

In other words, according to Andrews (2018) who responds to Buttlemann, apes may not be evaluating false beliefs of another but engaging in experiential logic outside the realm of making a proposition. This stance does not discredit Krupenye's or Buttlemenn's findings but simply suggests that apes don't need belief concepts to read minds. This is doubtful, though, since great apes clearly demonstrate self-knowledge and so understand their own judgments, mental images, and motivations. Like many people, an ape won't know the internal pieces of an engine; but an orangutan will know how to turn it on/off and what it's capable of doing. Any strict, hierarchical divide between ape/human must be questioned. Human language is not the pinnacle of communication, except for humans. Andrews (2017), for instance, takes Krupenye to another level by asserting that apes, especially chimpanzees, engage in "belief reasoning." This implies that beyond beliefs about objects, apes reason about perceptions, where such thinking carries implications for moral individualism. With theory of mind there are a number of mental states that could be attributed to another: emotions, images, physics, or needs. How an ape perceives or considers the beliefs of another in resource-abundant habitats impacts ecosystem engineering, that is, the movement in and manipulation of fruit trees and other foliage.

As an added kink in this debate, Caroline Spence et al. (2017) discuss what they call theory of animal mind. The question about how humans form beliefs about animal minds has ethical considerations, especially as empirical measures like fMRI (functional magnetic resonance imaging) reveal comparable mental states of consciousness in animals and humans. There are a number of factors leading to a theory of animal mind. For example, mechanisms in human thought leading to the attribution of beliefs in animals include simulation theory, where human models of thought are used to simulate animal thinking, and cognitive dissonance, a deliberate misalignment of beliefs. Simulation might be species oriented, in how humans see great apes as socially and mentally more similar to them than other animals (excepting their pets). There is less mental separation between humans/great apes than, for example, humans/cows or humans/pigs, partly explaining why many people justify eating animal flesh (but not their pets) in spite of cow and pig sentience and sapience. Another part of the explanation is that people don't see the animals slaughtered, only neat packages in supermarkets. This cognitive dissonance can vary across cultures, since some people live in closer proximity to animals than do most Westerners. There are multiple dimensions, as well as contexts, feeding into a human's belief about an animal's mind. One attributes more human tendencies to her pet than to a wild chimpanzee and even less to a primate behind bars in a research lab. Moral feeling has thus been expunged. Moreover, a lack of moral duty, in turn, allows humans to discredit the psychologically moral states of great apes and so take over their forest homes.

In spite of my harping on ape/human cognitive similarities here, let me return to a key difference. Evidencing their evolutionary longevity and survival skills, apes engage simulation theory with their forest habitats with little cognitive dissonance. They are not robots stomping through a forest but creators of and participants in the subtle manufacture of tropical biodiversity with positive impacts across earth's atmosphere. This land ethic is also reflected in the moral individualism of their social perspective taking since apes resource and share foods.

COGNITIVE COMMUNICATION WITH REFERENCE TO TOOLS

The logic of my argument now moves to how apes communicate with each other through gesture and implements; implied here is how they also "communicate" with their forest ecosystems employing their high degree of intelligence and theoretical thinking. This section is multilayered, so bear with me. Drawing from the high intelligence of apes just demonstrated, the import here is to show how apes combine communication and tools to engage socially in their forest habitats.

On the one hand, Anne Russon and David Begun (2004a) correctly assert that evolved ape gestural communication in the wild is symbolic. On the other hand, Sue Savage-Rumbaugh et al. (2000), stressing cultural consciousness and not only control situations, show how two apes or an ape and a human can communicate symbolically without gestures by using only lexigrams. Symbolic usage and advanced competency are present in apes, say Laurent Dubreuil and Sue Savage-Rumbaugh (2019). According to Julie Smith and Robert Mitchell (2012b), one does not need language to represent thoughts. Because language is semantic, linear, and based on subject-predicate, it is more confining than free-floating mental representations reacting to an environment. In other words, a look or a gesture can express meaning and emotion more quickly, at times, than a series of sentences.

Stephen Cowley and David Spurrett (2002) argue for the embodied brain in the world. Using the "languaged" bonobo Kanzi as an example, they don't necessarily see brain specialization (i.e., human uniqueness) in "language" but in how cognition employs the body in its environs. I suggest that this distributed cognitive communication is part of eco-engineering with tool use. Apes are not "languaged" in the wild, so their mastery of language, like Kanzi who was reared in a research facility, although with access to a nearby forest, shows their innate ability to communicate with objects. Indeed, Kanzi used lexigrams as other apes have used their hands for interspecies (human/ape) sign language, conspecific gestural interactions, and foliage manipulation.

Where most thinkers see, following Darwin, continuities between humans and great apes, others see disconnections. Derek Penn et al. (2008) claim human uniqueness by virtue of language, theory of mind, and symbolism. It seems disingenuous to deny by proclamation, in spite of the evidence, great ape mental abilities in communication, culture, or causal reasoning. Such a position is personhood denial. The key question is why any dramatic discontinuity between great apes and humans would evolve. Arrogance about human language does not obliterate ape cognitive or communicative abilities. There's vocal fold control or the ability to adjust pitch variations and frequencies, for example, in making vowel and consonant sounds. As it turns out, Adriano Lameira et al. (2016) show that there is ancestral vocal fold control in the orangutan's repertoire of vowel-like sound modulation. Human language is built from the vocal fold skills in great apes derived from a common ancestor. Furthermore, the rudimentary primate vocal tracts can easily produce five distinguishable vowel sounds, essential for phonetic production (Fitch et al., 2016).

Drawing from the extensive research on ape communication, which includes signing or lexigrams for orangutans, gorillas, chimpanzees, or bonobos, there's validity in Darwin's idea that human language is merely an extension of signals and symbols used by other animals. This theory seems to be borne out by the range of great ape species who, as H. Lyn Miles (1999) notes, communicate symbolically. Humans seem to communicate symbolically continuously (Tague, 2018). Is that merely a species difference? Because of vocal fold control orangutans can best imitate human speech compared with other apes. The FOXP2 gene is related to language and appears across species. Although this gene is conserved in most mammalian brains it was somehow accelerated in the *Homo* lineage (Staes et al., 2017). Work by Elizabeth Atkinson et al. (2018) shifts this paradigm, however. FOXP2 was not quickly altered in any sweeping manner among *Homo sapiens* to generate speech, since the code is found in the genome of Neanderthals, who split with modern humans over 600kya. There seems, rather, a derivative of the gene specific to *Homo sapiens* over other primates. This means the so-called language gene is not recently selected in humans. The FOXP2 gene is also connected to orofacial muscles, voluntary control that is apparent in ape vocalizations. Apes reveal variation in calls, probably inherited and modified from social transmission. For instance, says Staes, the long call of the male Sumatran and Bornean orangutans differs in terms of pulse, speed, duration, and frequency. A so-called fast call has also been identified (Galdikas & Insley, 1988), which is quicker in tempo, less common, occurs after contact or conflict, and elicited by other male long calls. How do we account for gene mutations and vocal differences between species and even among individuals? Most likely the answer is FOXP2 mutations, heritability, or social learning.

To what degree can one measure ape/human similarities, and does it matter? For instance, does a signaling chimpanzee understand another's mental state or simply respond to the other's behavior? Catherine Crockford et al. (2017) say chimpanzees are indeed aware of another's perspective, demonstrating that signaling and social cognition are evolutionarily very old and precede language. Chipping away at the tight definition of personhood, this means that there are instances of communication influenced by another's and not just one's own perspective or knowledge. Chimpanzees don't simply respond to another's physical behavior, such as callbacks or tree climbing, say these authors, but to what they think the receiver of a signal knows. With human experimenters, Crockford goes on, chimpanzees seeking food resources alter vocalizations in accordance with the attention, or not, of the human, showing how chimpanzees can even monitor awareness of a related species.

Similarly, cognitive communication and tool use are related to learning. Gisela Kaplan and Lesley Rogers (2000) report on research that shows how one brain region which increases in manipulation control subsequently influences expansion in other brain areas. This is not surprising since all brain areas work together in elastic sympathy. In this way, apes communicate cognitively through tool use and in learning how to use tools. I (2018) write about the coevolution of material/art in humans as a result of consciousness, communication, and cognitive culture. It has been determined that tool use in crows increases their optimism (McCoy et al., 2019). This important finding means that tool using animals are intrinsically motivated to engineer their environments, as I've been variously arguing throughout. Apparently, then, there are similarities (with some differences) in apes and other animals in terms of their mental interactions to survive and, consequently, sustain their environments.

For now, let me focus more on the communicative aspects of apes since, in total, I believe it speaks to their personhood. Evolutionary precursors to human language, argue Richard Byrne et al. (2017), can be found in the extensive stock of ape gestures. Indeed, it's not coincidental that the large repertoire overlaps across species, from orangutans, gorillas, chimpanzees, to bonobos. A possible common feature of human language and ape gesture is the compression of sequential sounds or hand movements, say Raphaela Heesen et al. (2019). A pattern of shorter units of communication is used more frequently than longer units; likewise, the more units there are overall, the smaller each one will be. This is not to say that apes speak as do humans. Rather, there is perhaps, suggests Heesen, a commonality in an underlying feature of communicative signaling. Nonetheless, like language, ape gestures, to be distinguished from signals, are deliberate, attention seeking, sequential, and aimed to influence another's thinking. In fact, according to Byrne, gestures can have distinct meanings made clear only in behavioral context. Though gestures are innate, like language they can be shaped and developed over time as they are used and shared across generations. Whereas monkey signals have adaptive, resource value, great ape gestures accompanied by vocalizations have goal-oriented intentions indicating awareness of a plan. For example, orangutans and chimpanzees monitor and judge the response of others to their gestures and will persist in order to achieve the intended outcome. Brain plasticity from learning and goal-oriented intelligence can bear on environmental manipulation.

Chloe Gonseth et al. (2017) say that ape referential gestures are language-like. Chimpanzees designate food resources at a distance by employing high manual gestures and larger mouth openings. Vocal flexibility to express emotions without depending on context is an evolutionary advance. Zanna Clay et al. (2015) show that as in prelinguistic infants, wild bonobos use a peep sound that is context independent and which accommodates positive, negative, and neutral emotions. Whoever hears these peeps, whether during travel, feeding, or grooming must make reasoned inferences about meaning. However, in a negative context, like alarm calls or aggression, the peep will exhibit different acoustics. In evolutionary terms a lack of vocal fixity is important since it means that very similar sounds are detached from functions and can, therefore, carry a variety of psychological states. Bonobos and chimpanzees express intentional gestures so similar that there is extensive overlap; theoretically, they might be able to understand each other, say Kirsty Graham et al. (2018).

There are approximately seventy distinct chimpanzee or bonobo gestures, and as with human language, an ape gesture can have more than one intended meaning. While the number might vary depending upon the expert consulted, the bare minimum number of human words needed to function linguistically is only several hundred. Kim Bard (1992) says that orangutans as young as only a few months old engage in intentional interactions with the mother. Intention in self-directed behavior is seen in the communicative function of loud scratches to coordinate mother-infant orangutan travel (Frölich et al., 2019). Chimpanzees also scratch to communicate in social contexts. Gestural communications come later in development, demonstrating object manipulation and social communication from a very early age and which grows exponentially over time, importantly, for my discussion, with experience and contingent on interactions with the environs.

Some researchers, however, do not believe there is meaning, as commonly understood, in ape communication. Thomas Scott-Phillips (2016) says that meaning is intentionality, where there is an intention in the sending of a signal so that the recipient grasps not only the intent but understands that there is significance in the communication. Scott-Phillips says he's only concerned with overt intentionality. The question, then, is whether or not there is ostension, the act or process of showing, in great ape communication. Ostension draws attention to a gesture indicating that something is being communicated, like eye gaze. Richard Moore (2016) believes this is so in apes. Josep Call and Michael Tomasello (1994) tested the signing orangutan Chantek and discovered that, perhaps because he was encultured by humans, he demonstrated better understanding of pointing than others. Tomasello says apes can only achieve high level cognition with human coaching. Dare I say that on some adapted level with little coaching apes supersede human intelligence when it comes to ecosystem engineering? Chantek's gestural communication and pointing included eye movement from the human he was directing to gazing at the target, meaning that he understood what the actors needed to do jointly. Gisela Kaplan and Lesley Rogers (2000) cite research that demonstrates how orangutans and chimpanzees understand pointing. Chantek did not simply copy signs but freely combined some and initiated intention.

The question of ostension hinges on forming mental representations and concepts as well as understanding inference. Ostensive behavior, like eye contact in humans and apes, expresses and assesses "communicative intent" (Gómez, 1996). As shown throughout my book, apes are capable of those higher order thinking abilities, especially in relation to their habitats. Apes do seem to look at other ape faces when acting, especially in defensive or offensive displays, as if intending to elicit a response. They know that for any gesture to be effective, the recipient must see it. A baby orangutan only several months old can respond to human faces pictured in a newspaper or magazine, Barbara Harrisson (1962) discovered. On a related note, Pamela Douglas and Liza Moscovice (2015) say that female bonobos use iconic gestures, which imply shared intentionality, to request genital rubbing. Despite the question of intentionality, apes psychologically communicate in and intend to physically impact on habitats, meaning they are socially defined by how they keep company with trees.

Loulis, Washoe's adopted son since her baby had died, culturally acquired sign language through her teaching so that eventually he could use sign combinations. Later, Loulis and Washoe participated with other signing chimpanzees so that Roger Fouts and Deborah Fouts (1993) could study intra-species communication. Loulis had a new companion, Dar, and much of his signing was about play. If there was a fight or ruckus, with Washoe running in as mediator, Loulis would blame his playmate Dar, all the while signing. Then Dar would sign to Washoe to hug him. For anyone who has watched human children playfully interacting, this behavior should sound familiar. Among the five signing chimpanzees in the Fouts study, about 88 percent of over 5,000 signing conversations dealt with play, social relations, or reassurances. Contrary to what might be assumed, food was not a major topic, unless they talked about foods they liked. (Of course they were provisioned with food.) In closed videotaping, the Fouts discovered that, remarkably, the chimpanzees talked to themselves, especially Washoe going back to when she was very young. Of the 5,000 chimpanzee conversations, nearly 120 were solitary, indicating their ability to think privately along with the ability to communicate and learn. In fact, in the many hours of remote videotaping, there was a high proportion of self-directed conversation with evidence that the apes were thinking about objects not present. These findings in communication not only support claims of great ape intelligence and personhood but point to the probable connection between learning, thinking, and tool use in ecosystem engineering.

There are parallels with Chantek, the signing orangutan, and Koko, the signing gorilla. H. Lyn White Miles (1993) says that Chantek was no mere imitator but an active initiator who combined signs to communicate. As noted, Chantek could point beginning at the age of two. Chantek's increased vocabulary enabled him to express ideas, and like the Washoe chimpanzee group, he signed about his own behavior using self-references. Similar to Washoe and Koko, Chantek could comprehend many words spoken in English. Once he heard someone on the radio meow and say cat, so he signed CAT. By two years of age Chantek had mental representations since he could sign for objects not visible and not in the vicinity, like signing for a car ride to get ice cream. Just as human children anthropomorphize their toys, Chantek would vitalize his toys in active play. Not surprisingly, Chantek was an avid tool user with complex interactions, such as painting, drawing, and food preparation. So he understood and planned outcomes in addition to being able to communicate. As expressed by H. Lyn Miles et al. (1996, p. 281) in relation to Chantek, but surely applicable to those like Koko, Washoe, and bonobo Kanzi, an ape can become an agent "in a meaningful system of relations that he would begin to embody as he acted within that system and as his understanding was affected by it." This apt quote describes a "moral" individual becoming a person in a community of others and a habitat using objects and signs as means of communication and resource feedbacks. Apes are intimately aware of themselves across multiple relationships, mental, physical, and ecological.

By her early twenties, Koko the signing gorilla had an impressive array of about 500 signs and a vocabulary of over 1,000 words, say Francine Patterson and Wendy Gordon (1993). Koko was bilingual, in that she responded with signs to verbal questions. Like Chantek and Washoe, Koko was self-aware in her activities before a mirror and engaged frequently in self-reference, consciously describing herself, her feelings, and her ideas. Koko could lie and also predict another's behavior. She played imaginatively alone or with another, and could produce representational drawings. She grasped the past and future as concepts. In addition to her large vocabulary, Koko could respond to several thousand spoken words. Koko lived with another gorilla, Michael, and Patterson and Gordon report how the two initiated conversations, created word combinations, and understood connotation, for example, how the word ROTTEN could refer to a person's behavior. She also understood dual meanings. When asked what is hard, Koko signed with the words ROCK and WORK. She even understood metaphor, using the sign RED to express anger. Frans de Waal (2019) relates how Washoe, when pestered by a macaque, signed DIRTY MONKEY. Indeed, Patterson reports that Koko and companion Michael had an acute comprehension of good and bad behavior like Washoe and Chantek, and so an understanding and expression of moral individualism. As with Washoe and company, there was also much self-signing. Koko and Michael taught each other signs, and Koko taught signs to her dolls, to the extent she'd mold their tiny hands.

As many of the last paragraphs in this section show, great apes engage in complex even cooperative communication. Human written and vocal language should not be the test of personhood. Apes have a "language" along with rational thought, mental representations, mind reading, problem solving, invention, willfulness, consciousness, conscience, and a concept of self. All of these attributes are important for learning and teaching, and combined with object manipulation provide forerunners to complex tool use in ecological self-sufficiency. Some might argue that apes are dull and lazy since they don't have sophisticated tools. But of course there is an easy rebuttal. Apes have the tools they need for survival in their environments. Was early *Homo* lazy by using the same toolkit for about one million years? Some others will say that the signing apes simply mimicked the human culture in which they were raised. There is some truth there, but it only makes the point that apes have an extraordinary capacity, beyond other animals, to learn another's culture. Humans assimilate into other cultures, too, and so there is a connection between cognitive capability, intelligence, communication, learning, and tool use or complex hand movements for food resourcing. A significant difference is that the cultural milieu in which apes thrive is a forest.

A key point I keep returning to is how ape culture is ecosystem oriented. Their moral individualism is expressed in small groups with relationships in and to the forest. Experiments in problem solving regarding food competition sometimes show that chimpanzees know what others have seen, report Brian Hare et al. (2001). Furthermore, there's evidence that the chimpanzees utilize such knowledge for strategic purposes. A chimpanzee could see fruit in a tree but refrain from looking so as not to alert others. There's understanding of when not to communicate. Chimpanzees can hide emotional expressions by putting their hands over their faces. Here the emphasis is on competition between chimpanzees rather than cooperation. A subordinate chimpanzee will go for food he knows a more dominant one is unaware of. That is not learned behavior, says Hare, but something more basic. Chimpanzees are aware of what others in the group know or have seen, more than simply following another's gaze.

There's the question of whether, like humans, apes can connect cognitive abilities of, say, tool use and sociality. Animals working from memory and reason can act with purpose toward a goal. Look for internet videos of Wolfgang Köhler's chimpanzees from the early twentieth century solving food resource problems by stacking crates to reach bananas, not including studies of the past fifty years. Just as "language" is a tool, physical implements can be used to signal intention or prompt cooperative behavior. Anne Russon and Biruté Galdikas (1995) conclude that imitation and tool use among orangutans grouped together with people at reintroduction centers are not wholly disconnected. Orangutans in these settings become bicultural in how they interact with conspecifics and then, watching humans, often randomly pick up a tool and simultaneously participate in a type of social communication with the implement. This shows how different modes of cognition in the ape brain can be related and put to use for resource gathering.

Planning for tool use, based on research by Nicholas Mulcahy et al. (2006), predates 14mya and evolved from an ancestor common to all great apes. Their research shows that orangutans and bonobos will select, transport, and save tools for future use, a higher cognitive skill since there is a delay between action and reward. Apes can apply a theory to an object, use it as a tool to test the theory, and furthermore use the manipulated object as a means for teaching or learning, all of which can involve cooperation. It's known that chimpanzees will carry tools from one location to another. Even a female chimpanzee will carry a heavy stone and her baby in order to crack nuts elsewhere. In addition to spontaneous, novel tool use, in an experimental situation, Mulcahy (2018) finds orangutan innovation of a stick tool for possible future use. This behavior indicates not just foresight because of a tool in hand, but anticipation of and preparation for a *what if* scenario with a tool that has been deliberately set aside visibly and ready for such an imagined purpose. The tool so conspicuously placed acts as mnemonic communication to the ape, and perhaps as an object to stimulate curiosity for others.

Thus, these apes in theorizing about object use consider future needs and not simply current ones, evident in many other species. In this respect there is mental time travel, once thought only a human capacity. Bonobos in a sanctuary have been observed to employ sticks and rocks as levers, pounders, or extractors, at times with dexterity (Roffman et al., 2015). Bolstering Mulcahy (2006), Mathias Osvath and Helena Osvath (2008) confirm that orangutans and chimpanzees engage in metacognition, or thinking about one's thoughts with consciousness similar to that of humans. Orangutans search for information metacognitively when they have scanty evidence (Mulcahy, 2016). These great apes control immediate desires and mentally construct events, through prospection and retrospection, for future tool use. Furthermore, great apes can take a tool now and generalize its potential usage for another task in another location while under the mental duress of maintaining ongoing social operations. The ability to detach oneself from present activity to consider a future goal is higher order thought processes. Critical to my discussion is how ape tool consciousness seems directly connected to eco-engineering. The ape's sustenance needs impact widespread ecological sustainability which appears to be deliberate since it occurs routinely with conscious intent.

For great apes, theirs is a language of gestures and so why the communication can be related to objects and tools with an ecological purpose. Anne Russon (2004a) set out to determine if orangutans can reason. She focused on imitative behaviors, many by ex-captive orangutans in centers that would ready them for forest reentry. The list of behaviors orangutans imitated from humans, without coaching, is impressive: sawing wood, writing, pulling weeds, sweeping, brushing teeth, washing clothes, boating, using blowguns, digging with a shovel, hammering, siphoning, pumping, painting, washing dishes, combing hair, using a parasol, using insect repellent, soap lathering. One even attempted to make a fire. Gisela Kaplan and Lesley Rogers (2000) relate the following story, told to them by Anne Russon. Orangutan Apollo Bob intently watched a Camp Leakey staff member relax by playing a guitar. When the staff member went inside his cabin for a moment, Apollo Bob filched the guitar and ran off with it to test his musical talents at the guest house. If this is not an example of an attempt to use a special "tool" for communication or at least personal expression, what is? It's well known, too, that apes enjoy painting creatively. In another story, one orangutan took a boat across the river to feed, and once there he held the rope of the boat, knowing it would float away if he did not maintain his grip. In this case, the shared community tool was cleverly used to succeed in food resourcing with the intention of future use.

Contrary to some conventional wisdom, orangutans are not only tool users but tool makers. Elizabeth Fox et al. (1999) studied orangutans in Suaq Balimbing, Sumatra, as tool users and found that tools are made from live twigs or branches, some transported to a fruit feeding site. Some tools are used for obtaining insects, honey, or for opening fruit husks to access and extract seeds. For insect feeding, stick tools are manipulated in a number of ways to achieve the goal of extraction, and sometimes more than one tool is used, that is, a newer or better one. To reach *Neesia* seeds, a tool is used to scrape away stinging hairs and then to push out seeds. All of this means that orangutans have a toolkit related to their eco-engineering. There are various tools manufactured differently, in terms of length and width, for either insect or seed extraction contexts. Tools are often modified. Previously, this was seen only in chimpanzees. Tool behavior transfer is likely where there is less distance between orangutans in social feeding parties or in male-female consortships. In terms of ontogeny, infants closely observe the movements of competent and frequent tool using mothers, and in rare cases Fox has observed innovation by the infants. The area of Sumatra under study is swampy with greater food diversity. Beyond any ecological explanation, Fox says that certainly tools emerged not only out of necessity but from orangutan high intelligence. There is, therefore, a learning component of tool use as a type of social communication for ecological purposes.

Hook tool making has been recorded in orangutans who spontaneously bend wires to access food. This special behavior was previously only seen in New Caledonian crows, say Isabelle Laumer et al. (2018), and the manipulation is difficult for a human child. Laumer says five adult male and female captive orangutans bent, that is, hooked, a wire chosen spontaneously over loose string. Later, the researchers noticed the orangutans making refinements to the tool. Considering the rarity of tools in the animal world, this is cognitively significant. In human history, hooks are a late innovation, for fishing and harpooning, coming as recently as 23–60kya. In this study, Laumer says the hooks were used to lift baskets containing food items from a tube. This behavior occurred previously in chimpanzees who saw a demonstration by humans. So this tool use goes beyond simply using a branch as a raking device in that the orangutans are able to mentally visualize a complete tool and project how to use it in obtaining resources.

With active improvement of the wire implement for better results, observed by others, there is an element of cognitive communication present. The researchers emphasize the spontaneity of the activity. However, in a related study, Laumer et al. (2019) examine the "relational complexity" in food sourcing decisions that involved using tools for a greater reward. With orangutans, as has been shown with monkeys and some birds, decision making can be affected by tool use, where a primate will delay immediate gratification by using a tool in a more profitable manner for better compensation later. In this way, especially with public display, the tool could be an extension of mind connecting into the surrounding ecology.

Imitation of how and why to use tools was a significant advance for humans, in contrast to some thinkers like Josep Call who don't see true imitation in nonhuman tool use. Anne Russon (1999), on the contrary, says great apes use tools generally with a clear understanding of cause and effect. Apes seem to recycle aspects of their own thinking for improved use or combinations. Importantly, tool usage is learned by imitation only in part, since there are underlying cognitive elements necessary for grasping what's going on and what it means ecologically, to say nothing of individual practice and innovation. While some fine points of imitation can be lost, for example, how an orangutan did not properly imitate the twist-screw motion to join two poles, Russon asserts that the assortment and excellence of tool manipulation by orangutans rivals that of chimpanzees, the most habitual tool users. A key theme here is that there are different cognitive functions for social interactions (i.e., simulative behaviors) and tool imitation, but both are closely related and manifest especially in terms of ecosystem engineering. This includes learned nest building techniques.

The salient points of this long section, drawing from ape cognition and intelligence generally, are many. Apes communicate often and effectively. Apes are cultural (to be addressed from a different angle in chapter 3). Apes do not act with mechanical rote but think. The hallmark for ape imitation is improvisation and flexibility if something goes wrong. Perceptive as always, Barbara Harrisson (1962, p. 168) says this especially about orangutans: "Their minds are not set to imitate, but to explore." Furthermore, these apes combine tools and tool steps, whether making a fire or making pancakes. Orangutan Princess, Russon (2004) says, since she had previously used pens, once took a burning ember, blew on it to create charcoal, and then used it to write. What could she have wanted to communicate symbolically with this tool? What does this effort reveal about her consciousness, moral individualism, and personhood? Not unrelatedly, Princess was a capable learner who had been taught sign language by Gary Shapiro (1985). There's no way to know another's mind fully, but as with ape painting, the effort and intent for autonomous expression is obviously there. In addition to the combination of vocalizations and hand gestures to communicate, great apes use a variety of "tools" in subtle ways to telegraph complex ideas and intentions. For apes, objects can become a means of social communication in an ecological context.

ENVIRONMENTAL PSYCHOLOGY AND ECOLOGICAL ENGINEERING

Considering the above sections on intellectual aptitude and communication, crucial to moral individualism, let me now turn to what might be the ultimate, adaptive "purpose" in great ape cognition and intelligence. As with other sections and my entire book, this will all take more than one paragraph to unfold. General intelligence, according to Laura Damerius et al. (2019), consists of reasoning, abstract thinking, along with ideation joined with behavioral and cognitive flexibility. Previously, general intelligence, or what has been referred to as psychometric *g*, was considered only a human trait, but it is apparent now in chimpanzees and orangutans. A key factor in psychometric *g* is orangutan curiosity, linked to memory, learning by association, spatial understanding, and grasping the concept of cause and effect. This general intelligence operates in apes as part of eco-engineering, evident in causally influenced behaviors from nest building to food sourcing.

Quantitative experiments across thirty-six species measured self-control in problem solving (MacLean et al., 2014). Diet variety and not necessarily group size among primates appears to have been a driver of self-control. A feeding ecology of knowing what to eat and where to find the best nutrition exercised a selection pressure on self-control (patience and timing) with the effect of increasing brain size and executive functions. A larger neocortex with its metabolic costs encouraged foraging for the better though widely dispersed fruits providing both rich dietary and increased cognitive ingredients. With fewer primates as reliable seed dispensers, there would be, as predicted by T. Trevor Caughlin et al. (2015), a negatively altered dynamic in the ecological structure of rainforests. Seed dispersal is critical for tree survival since, when aggregated with neighbors, trees have a diminished chance of full life growth. There seem to be correlations, then, among ape fruit eating, intelligence, and rainforest ecology. As Alejandro Estrada et al. (2018) declare, with primates in peril, so too the forests of Brazil, Madagascar, Indonesia, and the Democratic Republic of the Congo.

Mutualism occurs routinely in the rainforest, in cycles noted by biologist Junaidi Payne and collaborator Cede Prudente (2008). There are, as examples they note, fungi at roots that take energy but yet provide soil nutrients to trees. Mushrooms grow from the waste of wood-eating termites. Insects thrive off fallen fruit and then move on to pollinate trees and vegetation. While bacteria may kill a tree, once it falls to earth its corpse becomes home to newly growing plants and other small, even microscopic, creatures beginning new life processes. There are about 20,000 forest dependent plant species across Sumatra and Borneo, thus encompassing a huge interdependent network of life forms. There are various types of peat swamps, mountain ranges, montane forests, and river forests. The highest population of organisms is in swamps, fertile valleys, and fertile lowlands. Orangutans are a crucial part of and substantively contribute to the rainforest biome which is critical to the earth's climate sustainability, and their intelligence has evolved a land ethic in these environs in order to service themselves and hence the ecosystem.

Other species, like ungulates, will eat fallen leaves from orangutan feeding and then, in their process of foraging, turn the soil. Subsequently, their excrement, too, acts as yet another fertilizer. Organisms eat in a network. Horned beetles, for instance, feed on fallen fruit and break it open where, later, essential bacteria thrive and recycle nutrients, which consequently help moss, mushrooms, and ferns. High humidity on the ground in the Indonesian rainforest permits worms, leeches, and other organisms like amphibians and crustaceans literally to swim on leaves and contribute their share to the community ecology. All these actors, whether in the tree canopy or forest floor, fuel the powerful engines of rainforest climatology and hydrology which support forms of life across the globe.

There really is no way to explain the complex biomes of Indonesia in a few paragraphs. The point is that orangutans have evolved to be integral users of and responsible agents in their ecosystems while maintaining the health of the forests. Here, too, Esther Tarszisz et al. (2018) clearly show with quantifiable experiments and not mere observation because of careful tracking and use of seed mimics that orangutans engage in zoochory or mediated seed dispersal. Critical for plant communities is the viability of seeds to spread into new germinating sites away from parent plant competition. While female orangutans are active in such movement ecology, they are typically foraging in ranges of about a squared quarter of a mile each with some overlap. Male orangutans have wider roaming tendencies, probably to avoid encounters with aggressive males and to scout mating possibilities. Seeds can remain in an orangutan's gut up to five days before passing. Tarszisz found that seeds from orangutan feces, and especially seeds expectorated, actually have a higher rate of germination than seeds directly removed from the plant. Sister great ape species, according to Sue Taylor Parker (1999), diverged in diet according to body size and, in turn, ranging and social patterns were affected. The foraging habits and ranging movements of orangutans, without crowding each other, service the biome fauna for all species, suggesting that their evolved mental and physical behaviors are adaptations for ecosystem engineering. One can say so of beavers or earthworms, who operate on small scales, so why not orangutans, who operate on distributed arrays of various plants and trees?

Rowlands (2019) would support an extended, embodied cognition where an organism works physically as well as psychically in a defined locale since the structure of that space is informational to infer data and achieve a goal. Bodily processes reveal cognition aggregated in several brain areas for a function. Borrowing from Rowlands, I can formally describe such action, for example, an orangutan's feeding. But that description would not address the psychosomatic functions of the behaviors, an accretion of motives, needs, and desires to maintain bodily homeostasis and ecological symbiosis. Important here is how the orangutan intuits and sees her environment, not how we see an ape automating in the world. Even for the ape, the brain is less about storage and more about dynamic interconnections of neurons assembling contents for a purpose. Therefore, environmental psychology is an important consideration for this discussion, especially regarding ecosystem engineering outlined in chapter 1. The gist of the remainder of this chapter, but for the final paragraphs on gestalt theory, is indebted to the work of comparative psychologist Louise Barrett (2011) and tries to tie together the preceding discussion of cognition, communication, and tool use.

There does not have to be a "reason" for ape ecosystem engineering. Humans look for reasons where there simply might be an evolved instinct to be conscious on an autonoetic plane, that is, spatial and temporal awareness, of physical and mental self. In this type of consciousness an ape would become detached from the present, consider the past, so as to imagine events and actions in the future. Importantly, this consciousness occurs in large part because of one's body in nature. More precisely, great ape behavior in forests has been favored by natural selection since it benefits the entire ecological community. For example, orangutan roaming behaviors, as they lightly strip foliage, permit tree limbs to regenerate. Their buffet foraging, heavy movements, and daily nesting create forest holes of space and light encouraging new growth, especially for plant species crowded out. With knowledge of hundreds of types of plants, orangutans deliberately eat a variety of foods to balance nutrition, as if comprehending they'd deplete a tree's fruit easily. Russon (2004a) intimates that orangutans understand their rainforest habitats and the complex ecology in ways to help them successfully thrive, especially since they interact with many other species in the community.

Some will hurl accusations of anthropomorphism and say such thinking pretends that apes predict their own behavior. Meantime those very critics anthropomorphize their pets. Their children humanize toy animals. Advertisers anthropomorphize and vitalize a range of inanimate objects. The meat industry anthropomorphizes animals in advertisements, as if a smiling pig is happy to be slaughtered and eaten. As Charles Darwin demonstrates, whether from *On the Origin of Species*, *The Descent of Man*, or *The Expression of Emotions*, routinely close observation of behavior in the field is sometimes worth more than data analysis in a lab or on a computer. Skeptics would insist that there's a gap between describing a behavior and offering a sound biological explanation as to why it evolved.

In response, there's likely an evolutionary reason for anthropomorphism, Gordon Gallup (1985) might say, because one infers like states of another since one monitors one's own thoughts. Reference is made to chapter 1 which applied Tinbergen's framework for natural selection of behavior. Apes occupy, indeed have coevolved in, a certain ecological niche. Where it is accurate to say their cognitive behaviors reflect the ecology in which they evolved, it is even more precise to say, in line with Barrett (2011) and environmental psychology, that their bodies are almost inseparable from their lush, life-teeming environment and that very connection is even more paramount than any debatable cognitive functions. In fact, there might not be need for much cognitive function to move around in an environment. Consider the earthworm whose brain is connected to muscles and nerves for movement and sensory detection. While this notion seems to contradict most of what is said in this chapter (i.e., the focus on mental representation), the truth is that it rather offers a shift of emphasis away from the brain to the entire nervous system and body actively participating in the living world.

This means, according to Barrett (2011) and contrary to some others I've discussed, that complex behavior and sophisticated cognition are not necessarily linked, since that's a human assumption. Similarly, there is no separation between perception and cognition. The brain develops from the onset of birth and is dependent on and part of the environment from which it learns. Web building spiders, says Barrett, have poor eyesight, and so the web is as if a physical extension of the spider's body. By analogy, consider the orangutan who lives among tree canopies. The trees are a natural extension of the ape's body and, therefore, psychically conjoined to the ape's whole being and healthfulness. Mirror and motor neurons help a brain embody not only another's actions but to coordinate one's own actions in a densely physical and sometimes dangerous world.

What matters less is brain size, and what matters more is neuron connectivity. There's a false dichotomy between perception and cognition. Instead, perception is an active process, says Barrett (2011), and energizes behavioral flexibility. In contrast to any social brain hypothesis, natural selection could more easily have favored perception and bodily movements over advanced neural imaging. This supports the claim that ecosystem engineering does not have to be a high level cognitive behavior but rather a strategy selected for in resource continuation. Barrett notes how non-web spiders with small brains show flexible hunting behaviors. Likewise, she notes, the family of salticid spiders use perception and not planning to guide behavior. Animals don't see the world as humans do. Why don't apes stay in one confined locale, ravage it, and then move on? Their foraging behaviors are an instinct-based pattern of trimming and pruning the foliage without exploiting it completely. Apes are evolved to make circular, return movements in their range environments, compared with stationary plants. As previously suggested, apes can't be packed into limited national parks along with many other species and be expected to effectively service the smaller forest.

This all might seem obvious and merely descriptive, but Barrett's (2011) environmental psychology offers a simple explanation of how instincts (to eat) are not concretized genetic behaviors but can be modified (eat across a diverse range). Some instincts are behaviors to learn, like imprinting (my example), where a newly born infant attaches to the mother to learn coping skills. By analogy, apes imprint onto woodlands to learn how to survive. Learning is environmentally induced and not wholly instinctual or cognitive per se. Instinct and learning enable apes to know the forest as a recurring resource. Environments are dynamic and apes interact with the local ecology in a generative manner. Is this behavior knowledge? Is it inborn or learned? What matters is that ape foraging and resource oriented tool use occur in a sustainable way, evidencing self-organizing niche development.

Does a worm "know" that it needs soil to survive? Yes, because its evolved body tells it so, just as an ape's body knows it requires the tropical forest for safety, resources, and social interaction. In no way to minimize their intelligence, as emphasized in this chapter, recall nevertheless how an ape's psychology is really a sensing body in a lively, verdant environment replete with variegated foliage, fruits, sights, smells, tactile sensations, tastes, and sounds. Like Frans de Waal (2016b), Barrett (2011) speaks in terms of an animal's *Umwelt*. This means that in any given ecological community not all organisms equally perceive, experience, or respond to the same environmental factors. Rather, each organism has a specialized field of consciousness where its adaptive fitness is updated and restored by virtue of its entire nervous system, which includes and is not limited to the brain. The brain does not work, nor did it evolve, separately from its body or the body in its environment. Though I can't necessarily prove apes are consciously systematic ecosystem engineers with a land ethic, their *Umwelt* behavior certainly predicts it.

Perception and action are not separate but a connected process since for an ape there is an ecological approach to seeing its world, anticipating the next meal, and foraging for food. The world is neither a cognitive illusion for apes nor for humans. There is only the world in which one must act. Apes move in a world of bright light, dark shade, and shifting objects, and their movements provide information about the environment. There is less emphasis, in this approach, on mental representation than on bodily awareness in physical space. Only one who actively explores has access to the kind of physically perceptive data necessary for survival. Mental representations are not excluded, as evidenced in my book, but perception according to environmental psychology can be privileged, as I hope I've shown by opening and closing chapters 2 and 3 with this approach. Conscious awareness, fundamentally, does not need to include memory or thought. The brain does not hold a monopoly on an ape's impressions of the world. As Barrett (2011) suggests, there's information in the physical environment, and it does not make sense for natural selection to complicate the perception of such by having a separate reality constructed in the brain. Mental phenomena are connected to, because they are part of, the ape's natural world. Anticipating the next meal means provisioning for future allocations by resource ranging to ensure more tree fruiting.

Contrary to popular misconception the brain is not a computer. What is called cognition is more at physical activity in the world than computation on an imagined slate. In fact, as Barrett (2011) seems to suggest, with a non-computational model, representations and time sequencing are less needed. The environmental data an animal's brain manipulates are extended to its mind but yet in a dynamic relationship with its body. The brain, technically, does not hold all data since the body is part of the perceptive process, and so new impressions affect memory in waves of flux. The fruit and foliage environment of apes, too, is always in a state of change. Apes "know" all of this and somehow subconsciously calibrate their actions with the trees sustainably. If the trees thrive, they do so as well. In such a highly charged sensory environment, the ape responds to the conditional (active perception), not to the unconditional (mental representations). Of course emphasis in my book is on the mutuality of environmental psychology and ecosystem engineering. Apes do respond to the conditional (perceptive imaging) through their moral individualism in social situations, which, in turn, can be ecological.

Barrett (2011) herself draws from multiple sources of Andy Clark, a philosopher and cognitive scientist, so that mental representations seem to count, unless I'm bending her thinking to suit my argument. For example, how else would an ape handle a moral issue or reciprocal exchange transaction? Even so, these representations are not divorced from the ape's environment but buried deep in it. The ape *Umwelt* is not the same as that of an organism limited to ground movement eating the plethora of tiny bugs. Organisms live in their niches, small or large, not in their heads. Ape cognition is in direct and irrevocable physical contact with the natural world on multiple planes. On some level the ape is aware of and welcomes this connection. The ape's brain does not simply reflect back an environment but is an embodied actor controlling the jungle trees through her environmental psychology. This is an important point since any mental representations in an ape's mind are not stored there as museum pieces but are part of a fluid means to cooperate with the fruited greenery. Mental representations orient an ape to participate in its biome. As Barrett suggests, perception and action work so closely together in quick sequence including feedback to revised response that in many instances there is no real cognition.

The body and its parts know where they are in space without any thinking. While the word "engineering" is loaded with suggestions of intention and purpose, much of what apes do to maintain their ecosystems is non-cognitive, as I'm suggesting by analogy. Yet, the ape body moving and exploring in bristling, verdant spaces enacts a deep effect on thought processes so that an ape's consciousness, consisting of moral individualism and a land ethic, becomes part of the ecosystem. Without boundary, the ape environment is part of their cognitive makeup. Apes don't destroy their habitats because the sustainable maintenance of the foliage and fruits is a mnemonic extension of their ability to access these resources. The upshot of this section is that great apes, via environmental psychology as ecosystem engineering, reflect a holistic regard for nature. They evince a moral individualism and land ethic. Why? The answer has something to do with how apes ignore the human market system outlined in chapter 1, where short-gain economic acquisition trumps long-term ecological health.

Because of paucity and wide distribution, food resources strain ape populations. Long intervals of birthing, especially in the orangutan, might have evolved as adaptations to mitigate, in line with Malthusian theory, a destruction of offspring and future generations because of food loss. Though orangutans like to eat, as there is a consuming instinct even among humans, a balance obviously exists in apes. Could there have been a coevolutionary pressure of population equilibrium and spaced food resourcing? Do apes sense or "calculate," in a manner similar to inclusive fitness, how the forest is like kin carrying their genes to successive generations by virtue of their eco-activity? Consider, in contrast, how expanding human populations and their typically voracious consumption are driving a multipronged crisis in terms of climate, water, and food. Darwin (1859, chapter 3) saw how adverse climate can have a profound effect on a creature's struggle for existence since food supply is then often choked. Nevertheless, fruiting in Indonesia is more irregular than in African tropical forests so that, according to Juichi Yamagiwa (2004), food resourcing problems for large-bodied apes necessitated social and cognitive advances as outlined in this book. In turn, like a ratchet effect, once some obstacles were overcome in the environment of evolutionary adaptation, cognitive advances were evolved to tackle other difficulties, like communicative tool use and eco-engineering.

GOOD GESTALT

As a final word in this chapter, it would be instructive to look at the differences between great apes and humans through gestalt theory, especially as it relates to environmental psychology. Gestalt is a unified configuration where the overall pattern is not merely a summary of its parts. In gestalt, a part affects the whole and the whole affects the parts with a sense of belonging together. For apes, we can see how the physical environment is unified, to which they are integral users and communicators. Gestalt is critical for perceptions of all kinds, whether sight, sound, smell, or touch, and as a person of the forest, for instance, an orangutan is not there randomly but fortuitously. Most modern humans, on the other hand, have managed to separate themselves, literally and figuratively, from "nature." The key difference in terms of apes/humans is that apes seem better at achieving good gestalt.

There are three related phenomena to be conserved in relation to gestalt as derived from Margaret Korb et al. (1989). There is the perceiving object, its environment, and how both relate to each other in a creative process that integrates a whole. Gestalt is the experienced whole from parts in context, how all pieces function to make meaning for each other. There's the individual who's self-aware in relation to her awareness of the environment.

Gestalt is a pattern maker, a natural process where a sense of the present is paramount overshadowing, but not excluding, past or future concerns. This means there's existence of clear perception in the whole environment for good gestalt. Total awareness is crucial for good health, to be distinguished from attention, which is directional and which can impede awareness. Clearly no explanation is needed to understand, here, the difference between a human and his perceived needs in a constructed environment and an orangutan and her real needs in a rainforest. Which of these two beings is physically and mentally healthier?

This good gestalt is not a will to power, not a force. Key to gestalt is holism, but not a mere accumulation of bits that make the whole. In holism, there are no isolated parts of the psyche. This is important since with gestalt, there's less emphasis on the linear, analytical, and rational and more to intuition. Evident in my book and contrary to Hegel, the real is not the rational. Western human achievement hinges on reason and wants to separate itself from real experience (an ape's forest world). Here, then, is a key difference between the argument in my book and others claiming ape personhood. Certainly this book shows how apes are rational and engage in, for example, complex thought regarding intention, tool use, and theory of mind. But the critical difference is that, leaning toward environmental psychology, *An Ape Ethic* tries to downplay too much emphasis on rationality. That extra weight on rationality stems from Kant and is used as a hammer by courts against sentient and sapient beings in any bid for forest personhood. Like a city kid who is "street smart," apes are environmentally intuitive and communicative, in a gestalt-like way, in addition to their rational functions which evolved as ecological adaptations.

Yet as one might imagine, there's a significant difference between city smart and forest clever. With good gestalt the person is not divided into parts. Such division often happens when too much stress is placed on rational processes separated from experience. Bad gestalt is a willing suppression of sensory awareness. The ape's care for her offspring, feeding, socializing, and nesting are all one as part of a lively ecosystem. There are no separate spheres. The examples Korb et al. (1989) give of poor gestalt include city living (noise, crowds, pollution); compare that with the forest habitat of apes and how they, in contrast, engage in heightened awareness. On the contrary, city and even suburban dwellers forcibly darken sensations of their physical surroundings. Think of all those lawnmowers howling on Saturday mornings. Curtains are drawn; windows are closed. Though my book explains great ape intelligence, thinking does not cause the physical environment. The whole experience of living is not the rational but, rather, a more vital and visceral bodily process. Perhaps these ideas, related to *Umwelt* as well, are in line with Gaia and how all organisms are in a fluid totality of change, an integrative consciousness. This interaction implies a natural and not rational (artificial) ethics. The good gestalt organism is self-regulating toward wholeness, which is not necessarily dependent on a rational procedure.

The gestalt formation typically follows this uninterrupted cycle, following Korb et al. (1989). From rest, there's sensation for need. This is followed by awareness of the environment, movement and choices, direct action and contact, and acceptance or not. Then, there's reduction or satisfaction of the requirement and release to preparation for new necessity awareness. It's far easier to imagine this happening to an ape in her habitat than for a modern human being.

Important, here, is how excessive cognitive control, seen in the case of humans, diminishes or frustrates natural abilities for an organism to stay in balance with its environment. In line with evolutionary psychology, an infant orangutan has inborn psychological modules and intelligences to survive in the forest but, nonetheless, needs its mother to shape and help reveal forest skills successfully. Not all cognition is in the brain, Shaun Gallagher (2014b) might suggest, adding Heidegger's notion of *Zuhanden* or the posture of being ready-to-hand. Humans tend to situate themselves in the world via mental possessions and forget that they are embodied organisms embedded in a physical and social environment. As I hope I've shown, great apes are very much in tune with the bio-network they inhabit and to which they contribute. Incidentally, cognitive control praised by rationalists is not necessarily moral since it can quash empathic, sympathetic, and other-directed natural impulses, emotions, and feelings. For instance, humans have rationalized reasons to farm and slaughter animals, to degrade water tables and land by mining, and to deplete ocean resources.

The self is manifested when in contact with the environment, which can but does not have to include other persons. A template for good gestalt, say Korb et al. (1989), appears as follows:

1. Experiencing the here and now, immediacy;
2. Self-acceptance;
3. Interaction with the actual and not wished-for environment;
4. Self-honesty;
5. Avoiding self or interpersonal manipulation;
6. Accepting and expressing all emotions;
7. Willingness for experimentation and change.

Interestingly, like humans apes are susceptible to failure in numbers 5 and 7. All in all, responding to the environment is akin to responsibility, an act of understanding one's limitation and managing expectations. So from the list, and observable in reality, it seems that great apes are more in tune with good gestalt than most humans. Let me offer an example. As opposed to apes as eco-engineers, humans transform ecosystems only for themselves. Rather than a strict struggle for existence in competition, or what Herbert Spencer falsely calls survival of the fittest, Eileen Crist (2019, p. 221) says life forms are much more to "coexistence, interdependence . . . diversity . . . and varied relationships." Based on what I've offered so far, that rings true of great apes in their biological networks.

Neuroscientist and philosopher Francisco Varela (1992) offers a critique of over emphasizing representation in cognition. Importantly, however, as I have tried to do here, Varela suggests a middle course where cognition is understood as a process of living in the world to build perceptions. This means that species use bodies and brains to define their worlds differently. I know that may seem obvious, but it's critical to restate in an argument for apes as eco-engineers. There's a question of what's viable to the species for survival success across generations. Cognition is not simply a solution to any single problem, Varela might say, but has arisen among interconnected bodies and distributed minds in specific environments and not through passive mirroring alone. In this way, decisions are made about survival, reproduction, parenting, housing, food sourcing, and socialization—through actions more than simply representations. Nonetheless, as apparent in my account, I do not wholly discredit representations in ape mentality, moral behavior, and their land ethic. I cannot, since the ape mind is a version of the human mind, with the difference that theirs is geared to ecological sustainability.

According to Varela, then, and to other thinkers of embodied cognition like Shaun Gallagher, great apes don't think apart from the world but are mentally engaged in their physical surroundings. Clearly that behavior is unlike the anti-nature industrial activity we see in many modern human societies. Embodied cognition (over representation) for an individual might look like this, following Varela (1992), and which can serve as an apropos end to chapter 2 and a link to chapter 3. One moves from task to solution, from center to distribution, from representation to action, from design to adaptation. That's mostly physical embodiment and not only mental abstraction.

If cognition ushers meaning from activity, then knowledge is value and has "moral" implications. With that thought I now move to examine the remaining gears of this little engine of ape psychology in terms of their social and cultural behaviors related to eco-engineering. As a recapitulation, please remember what has been covered in this complicated chapter 2. I looked at emotional understanding in apes, their brain and consciousness, intelligence and instinct, their high level of thinking and theory of mind, communication and tool use, and, especially, their environmental psychology and ecological engineering of good gestalt. Knowing these topics is essential to understanding the complex eco-sociality of apes.

Chapter 3

Social Behavior and Personhood

This challenging chapter attempts to show how ape intelligence is connected to social activity in forest habitats. Emphasis will be placed on their "moral" capacities as individuals and in groups centered in a complex ecology. How a moral individual deals with others regarding resources is a hallmark of development. The biology of morality is connected to mutualism, and the ape ethic is tied to forest personhood or the recognition of moral individuals in an environmental locale with specific and at times limited resources. As primatologists Caroline Schuppli and Carel van Schaik (2019) say, human culture is normative, that is, sanctioned by social groups, but it stems from ape capacities that are not always but can be normative. Readers should note that where some emphasis, in chapter 2, was on ecosystem engineering, in this chapter the question of personhood via ape social behaviors will be underscored.

As with the few guiding words about enactive cognition I offered at the beginning of chapter 2, Shaun Gallagher (2014b) is helpful here as well. He reminds us, perhaps more apt for apes than humans, that situation is not simply environment but the organism in its ecosystem. One moves in and with a situation, Gallagher might say. Edwin Hutchins (2008) confirms there's a dynamic interaction between a body and objects in its culture, and this is apparent among apes and their tools. The ape's body animates stones and sticks with culturally communicative tool use for ecologically amenable food resourcing. There is a rearrangement and adjustment of the self, typically social, in relation to vegetation. Akin to extended mind, one finds world meaning through communal interactions. This would be true, too, of orangutans who, though semi-solitary, are not alone.

More particularly for this chapter, especially, Gallagher and Robb Lindgren (2015) relate how learning comes with full-bodied engagement. We don't necessarily teach our children in this way, but apes do. Humans might, instead, use enactive metaphors. An active approach to embodied cognition would include the following, as per Gallagher and Lindgren, and I hope readers will see, based on everything I've written so far, how all of these apply to ape as well as human persons. The difference is that some, like numbers 1, 3, 5, and 6, are perhaps more heightened in great apes:

1. Cognition is a distributed process of the body in an environment;
2. Meaning and intentionality are derived, in part, from actions;
3. The context of action should predominate over internalizing the world;
4. Enactive cognition is intersubjective and socially extended;
5. Learning should be a dynamic interplay of brain, body, and environment;
6. Cognition is in sensorimotor skills, emotions and bodily functions in an environment;
7. Reflection is linked to practical knowledge of an environmentally situated body in action.

As will be seen, these psychological routes occur in apes, so I ask readers to keep them in mind as the following sections unfurl and wrap up. It's well known that human children who use their hands in math problems score well, and being cognitively and affectively mindful of one's physical and social movements is apparent, too, in great apes. As I hope to show, building from chapter 2, tools and gestures as communication are part of a cognitive social event (Gallagher, 2009). Mental experience is part of the ape organism acting in a living region of plants and conspecifics which literally and figuratively feed understanding and knowledge.

Social organization among primates varies. Some are semi-solitary, with either or both males and females ranging, like orangutans. Some might pair, where males and females inhabit the same territory regardless of breeding, like gibbons. Under one male or more, some females bond in a loose group and travel together, like chimpanzees. Concerning social behavior and organization, Jerram Brown (1975) says there are three main drivers: agonistic (individual selection); mating (sexual selection); and altruism (kin selection).

All three of these point to how there is a concept of individual, private space among animals. After all, the individual is the gene container and unit of selection, so survival and successful reproduction are vital. Nevertheless, among primates these gene containers work best if they cooperate socially with others. I've extended this to note, repeatedly, that the forest is naturally "related" to apes as part of their inclusive fitness or how the properties of their actions are causally effective in habitats for their own health and genetic longevity. Taking a slightly different perspective, Richard Prum (2017) says that female mate choice in chimpanzees, bonobos, and especially humans, but less so in greatly dimorphic species like gorillas and orangutans, could have affected male sexual and social behavior.

Picking up from Susan Essock-Vitale and Robert Seyfarth (1987), one can say that great apes and humans share, on a fundamental level, not only intelligence as just discussed in chapter 2 but also a range of social behaviors considered here. Apes, and even some monkeys, possess a cross modal perception where through, for example, vision, they can grasp data that was previously experienced as touch. This advanced cognitive ability enables one to learn how to acquire social and environmental information and can be applied to interactive relational situations. For instance, self-conceptualization is of key significance and connected to an ability of using social stimuli across interpersonal interactions on a local and group level.

How are big brains related to social behavior? Some great apes tend to organize in small groups, like chimpanzees and bonobos, so a social intelligence hypothesis might not explain brain size. Nevertheless, as Robert Boyd and Joan Silk (2009) explain, there are selection correlations between big brains and long lives, since energy used for the brain benefits an individual's life. There are at least three explanations for big brains, and some of this explanation draws from the previous chapter and feeds into this chapter.

1. Social intelligence. This includes competition, bonds, coalitions, networks, and resource allocations. There is flexible behavior in a complex social world, with learning and planning.
2. Ecological intelligence. One needs to evaluate food resources, ripeness of fruits, and consider nutrition and toxicity. There is memory, learning, and planning here, too. Extracted foods generate protein, like ant/termite fishing or nut cracking, both of which require bodily coordination.
3. Behavioral flexibility. This is not just planning but inventiveness and innovation to challenges. There is learning from others and planning to help the group member socialize.

The bottom line is that while some of these three behaviors occur on an isolated, individual level, they can take place in and are part of a communal network. Primates are cognitive social beings with heightened consciousness of their actions, whether actual or anticipated, in their relation to others and to the forest environment. This self/other awareness is true even if contact is not on a diurnal basis. In one study, nonaggressive social interactions of dynamic exchanges among lowland gorillas, say Giovanni Forcina et al. (2019), foster culture and cohesiveness. The biological and social networking is, in part, related to the ecology and selected by the environment of evolutionary adaptation. Amazingly, transient, non-kin individuals, says Forcina, are tolerated and even accepted by others. As Darwin (1871, chapter 4) was able to see, there's a connection between self-awareness and moral behavior. Great apes are keenly aware of themselves and others, but cognizance is well beyond recognition and carries the seeds of "morality" in how they act. Importantly for my argument, ape social consciousness is evolutionarily shaped by a forest environment, where behavior is part of the ecology. Genes respond to physical and social surroundings, and there are many genes implicated in interpersonal interactions.

BIOLOGY OF MORAL SYSTEMS

This section borrows its heading from an important book by biologist Richard Alexander (1987) called *The Biology of Moral Systems*. Essentially, Alexander points out how methods of social behavior in the human realm genetically evolved to help, at first, offspring and kin, and then later, members of a group and a community. Do apes exhibit what we call moral behavior? Yes, since social knowledge is conceptual, categorical, and propositional. Raphaël Chalmeau et al. (1997) demonstrate that a pair of orangutans with no prior training cooperated socially in a pulling test to get food. Synchronized and simultaneous cooperation might be evolutionarily very old, since this behavior is not generally observed in orangutans who are semi-solitary. Social cognition is involved in this effort as well, since both recognized the notion of mutual aid. Though the reward among apes is not usually shared equally since there is comprehension of status differentiation, equitable sharing that is typically human has its roots in joint problem solving.

For instance, chimpanzees and capuchin monkeys, like orangutans, exhibit what is known as the endowment effect, say Timothy Flemming et al. (2012). This means that one ascribes more value to something simply because it's in possession, a divestiture aversion. There is a status quo bias in this behavior where one tries to mitigate loss of goods. This quasi-irrational conduct impedes transactions and on its face is antisocial. The exchange partner could defect, resulting in loss, and why in modern human societies there are complex trade agreements. Yet we know that apes share resources and exhibit empathy and kindness beyond kin, so they are willing to risk cost. What appeared as an adaptive behavior came to be used as a social strategy. Contrary to popular belief, orangutans are not without social bonds. Males have dominance hierarchies, though no real evidence of coalition forming or cooperative hunting. Food sharing occurs between mother and infant. While designated as semi-solitary, there can be some very small groups during food abundance, says Flemming. Does the endowment effect carry social or even moral implications? It appears to be a primate homology, says Flemming, and is evident among humans. In some cases it might simply be an adaptive advantage when in foraging one retains what's in hand during lean times in spite of an apparent favored item at a distance. That's a selection pressure on the individual level, what Richard Dawkins (2006) has famously called the selfish gene and from which human moral systems slowly evolved. In other words, for many social mammals individuals knowingly act in a group for protection, resources, or so that culture and, in the case of apes, eco-engineering, can evolve.

To begin, I'll take a quick glance at some relevant philosophical ideas that impinge on this part of my discussion dealing with social behaviors but that impact on the entire argument of ape moral individualism in their land ethic.

In a classic text on philosophy, Paul Edwards (1973) addresses determinism, free will, and moral responsibility. Can moral responsibility be justified if determinism is true? Like humans, great apes do not act on instinct alone, though choices are based not only on their own but on another's past behavior. While momentary circumstances seem to dictate action, the brain is wired to calculate and make predictions based on experience. How, then, does one act freely? A large part of the answer rests on the character and temperament of the individual, whether ape or human. The signing orangutan Chantek, according to H. Lyn White Miles (1993, 1994), could mentally represent himself in terms of others, so as to take perspective and deceive or empathize. That's a moral behavior. Chantek could understand and thus sign GOOD/BAD appropriately in reference to himself and others, especially self-signing about his behavior upon reflection. Koko the signing gorilla had a sense of humor and other social emotions, particularly empathy for those close to her, including pets, and even those in pictures (Patterson & Gordon, 1993). While there are species-typical or common cultural responses to social or environmental situations, individuals can behave differently according to assumptions, bias, or serendipity. No outcome is wholly knowable. But one can act freely by exercising reason over desire or instinct, by his or her own efforts to alter a chain of events or anticipated outcome. There is a power of individual will at work across species, evident, sadly, among those infant apes who abandon the will to live upon losing their mothers or when separated from their group by trafficking, as in the story of Jimmy (Tague, 2019).

It is precisely because in complex social groups behaviors are unpredictable that there is a moral responsibility implicated in individual will. A moral judgment is not a determined fact. While male chimpanzees patrol their territories as a given, it's a huge leap on their part to organize and systematically eradicate another group (Goodall, 1986). Is this a case where individual moral responsibility is relinquished to or determined by the group? The fact that two great apes can differ about how to act in hierarchy disputes by taking sides or forming coalitions points to moral disagreement and thus individual responsibility. Group members are aware that their actions will have consequences, seen in their expression of emotions, from joy or sadness to shame or guilt. Sexual, and even status, jealousy is an evolved adaptation. In turn, the expression of emotions can influence the actions of another.

Philosopher Richard Norman (1998) notes how W. D. Ross posits obligatory relations between persons as a prima facie duty above benefits for everyone. This is an ethics of moral individualism in social relations not limited to the human domain. Like eighteenth-century philosopher David Hume, one can say that moral approval/disapproval comes from sentiments where reason, later, only helps one understand consequences. James Rachels (1998) goes as far as saying that Hume's thinking suggests how moral sentiments can excite passions. Similar to many humans, ape persons are more to individualists and less to consequentialists, for the utilitarian sometimes fails to examine interpersonal relations and looks only to the abstract larger whole. Rather, a focus could rely less on childish rules and more on mature states of mind, as Bertrand Russell would say (Rachels, 1998, p. 19). It was established in the previous chapter that apes act intelligently and utilize theory of mind. Apes and other animals are not without sympathetic feelings for others, and this too can be a source of or guide for moral evaluation and action. This is not to say there are no benefit-oriented, group-wide *mores*, even among the semi-solitary orangutan where a "group" is small or widespread. Let me not, however, be simplistic by reducing ape motivation to the individual or kin level exclusively.

Many philosophers seem to acknowledge moral status for animals but not necessarily moral agency. The argument here is not simply to ask for moral status, since that consideration should be a given seeing how animals feel pain and exercise some high order cognitive functions. Rather, the claim is to assert that by virtue of their intelligence, complex social structure, and ecosystem engineering apes evince moral states of mind. They do act, as humans might say, morally. They might act, as humans often do, immorally. Apes don't need Platonic or religious principles to be moral, says Paul Shapiro (2006). It's a question of approval/disapproval of some actions, a question of helping/harming or understanding the difference between a mine/yours rationality tied to emotions. That's all human, too. Animals can make moral choices. If an orphan ape is taken in by a non-kin ape, the explanation for that caring behavior is not much different than a human adoption. Liz Campbell (2019) tells the story of a three-month-old male macaque separated from its group, injured, and in distress after a car accident. The juvenile was found two days later by another group who cared for him over the course of four months, at which time he returned to his natal group. This altruistic activity demonstrates affiliative behaviors and consolation in wild monkeys, our primate relatives from deep prehistory.

Similarly, are hunger strikes only human? No, since a macaque monkey will decline feeding itself if in doing so he limits harm to an unrelated conspecific. That's a "moral" choice, not to inflict harm onto another by an electric shock even if it means you can't eat (Masserman et al., 1964). Choice about care for others exhibits moral behavior, and in fact without principle. Consider how many humans might act kindly toward others out of principle (the Ten Commandments) but who personally hate doing so. In fact, early *Homo* species most likely had no principles but acted intentionally like the vast majority of macaque monkeys who, out of reasoned feelings and not instinctual fear reflex, refused food to prevent another's injury. Some reactions are ethical and come without moral deliberation, anyway, like saving non-kin from drowning without any expected reward, cases of which have occurred among apes (e.g., Fouts & Fouts, 1993; de Waal, 2006). Apes and other animals understand and willfully agree to social obligations, a key point.

Frans de Waal might emphasize how many animals, but especially primates and particularly great apes, engage in empathy. Environmental and animal philosopher Elisa Aaltola (2015) tells us there are four types of empathy:

1. Cognitive (mind reading)
2. Affective (echoing fellow feeling)
3. Projective (simulation of feeling into another)
4. Embodied (physical intersubjectivity)

Numbers 1 and 2 are important in terms of moral agency or feelings directed toward another being. As I've been arguing, numbers 3 and 4, with qualifications, relate to how apes cooperate with their ecosystems as well as to conspecifics. Many animals, and especially great apes, demonstrate all of these empathic capacities. Regarding another individual, empathy is more to identification and sympathy is more to active concern. Darwin would have stressed sympathy and how, through his reading of Adam Smith (1759, *Moral Sentiments*) there is a self-interested component to sympathy (Tague, 2014, part one). In helping another, perhaps there will be some type of physical or social status reward, a consideration that does not even need to be conscious. Elements of what most humans call morality can be found in primate social instinct, which, though grounded in self-interest, can rise to a level of empathy or sympathy with their habitats. Morality is in part a cultural creation that stems directly from the emotional behaviors, whether dominance or sharing, seen in apes, with whom we share common ancestry. While male chimpanzees will argue over a main meal, says de Waal (1982), they can be, otherwise, generous with extras, except among rivals. This is a form of control, since sharing earns respect.

Sharing can be public, reciprocity not necessarily so. Cor van der Weele (2011) has written about this subject and the nuance between the morally good and self-interest. Aggression and competition factor into moral behavior, for we cannot know what is good (e.g., altruism) without knowing what is bad (e.g., selfish greed). I know what you need; will I help you? Great apes, too, know this distinction. In an extended field study, Christophe Boesch et al. (2010) report on eighteen cases of known adoptions among wild chimpanzees. Half the adoptive parents are male, with only one the birth father. Altruism in this scenario is expensive where one offers extended care and welfare for non-kin. There is no genetic relationship and hence no genetic continuity. There is no guarantee of any future reciprocity, though that might be anticipated. The cases in this study demonstrate a high degree of in-group solidarity and nonexclusive sharing of care. Jane Goodall (1993) also talks about a case where a young male chimpanzee adopted a male infant whose mother had died. The young adult shared food, carried the infant, and slept with him. Goodall uses this touchstone example as a way to show how great ape moral emotions are not just similar to, but could be nearly identical with, human emotions.

Within their own community, ape members typically do not harm each other. They show restraint. There is a common evolutionary history for ape/human self-control via cognitive bases of general intelligence. Moreover, there's a relation of general intelligence to efficiency, which appears high in chimpanzees as in humans (Beran, 2018). Though crucial, this development is less to inhibition and more to persistent social cognition, patience or delay maintenance, and cognitive monitoring (Beran, 2014, 2018; Eisenreich, 2018). One can recognize sexual dimorphism, especially pronounced with gorillas and orangutans, which indicates classic sexual selection with differential reproductive success. Prehistoric male-male combat is inferred. Male orangutan bimaturism is a special case of dimorphism where the onset of large body size and cheek pad/throat pouch development is delayed, thereby minimizing male-male aggression until the right age and psychological conditions are met. Chimpanzees, like many humans, freely demonstrate aggression. However, a male chimpanzee who gets extensive maternal attention will be socially successful later. Orangutans, on the other hand, tend to avoid conflict in their thoughtful patience. Orangutans do not attack to kill, as some humans or chimpanzees might, and there's no history of any orangutan killing a human, says primatologist Leif Cocks (2016).

This is not to say an orangutan cannot be demonstrative, and orangutan aggression is not limited to males, as females can attack each other when asserting their dominance positions. The point here, though, is that with aggression comes a sense of knowing, across great ape species, when another is harmed. In some species, for example, chimpanzees, conciliatory behavior after conflict is common. Aggression implies a range of both self-oriented and other-directed emotions. An orangutan who directs aggression toward another knows what he's doing and understands that he could receive similar aggression. This level of intent and knowledge means the orangutan can distinguish between hurt-him/I-safe, a "moral" stance that implies help-him/I-give. The opposite of harm is aid (or at least indifference), and even a human infant registers that social fact. In terms of empathy, facial mimicry is a way to measure emotional contagion. In experiments by Marina Davila Ross et al. (2007), it was discovered that orangutans rapidly mimic each other in social play, evincing positive emotions akin to human laughter. This means that a cornerstone of human empathy or positive emotional contagion can be found in great ape homologies.

Whereas empathy is automatic and passive, sympathy can involve action. De Waal (2009) suggests that popular culture has presented apes wrongly, negatively as clowns or worse as beasts, and ignores the biologically programmed empathy of apes that is within humans, too. He goes on to talk about contemporary hunter-gatherers who don't fight, share water with strangers, and permit out-of-group marriage. This last example demonstrates external kin and so a disincentive to kill outsiders. This means, de Waal says, that any behavior deriving from evolution, whether natural selection or sexual selection, does not preclude an agent from acting otherwise. Selection pressures aim to benefit an individual or kin, but often mammals, primates, apes, and humans act in ways to help others with no discernible benefit. Borrowing from Darwin (1871, chapter 4), van der Weele (2011) says that sympathy by itself is not "moral" since morality involves making a judgment based on past experiences and anticipated action in light of approval/disapproval. Apes are capable of this, evident from findings discussed in the previous chapter and in the reconciliation behaviors of chimpanzees and bonobos. In Adam Smith's (1759) view, the feeling of sympathy becomes a commodity to be exchanged, clear in apes as well as humans. Apes, in fact, judge actions and merit approval, disapproval, and even punishment accordingly.

There were experiments done in the 1960s where rats and rhesus monkeys were electrically shocked near conspecifics. Given the opportunity a rat, and especially a monkey, will help or ameliorate the distress of another (Wechkin et al., 1964). Human infants, too, will respond to the distress of another, and as they get older they will spontaneously help others. Where do these tendencies come from, especially considering their appearance across species? That's a question Stephanie Preston (2002) posed, and she found her answer in empathy. How one perceives the condition of another is rooted in evolutionarily old mechanisms like emotions, attitudes of approach/withdraw, and parental/kin responses. As the German etymology suggests, empathy is a feeling into, where, according to Preston, the process of emotional cognition (sympathy and reason) factors into and increases aid in degrees depending on the relation of the subjects.

One can respond *with* (i.e., feeling the same emotion) or respond *to* (i.e., instrumentally acting). These factors are heightened in species who live in groups, like great apes and even considering the semi-solitary orangutan who is continually aware of the whereabouts of others. The social structure of orangutans evolved to be semi-solitary where individuals are visibly or audibly connected but given latitude and space. In spite of the seeming automaticity of any helping behavior, there might be, even on a semiconscious level, a quick cost-benefit analysis, as we'd expect in group living primates, humans included. Adaptive emotional connections and evaluations begin early in life via the mother-infant bond and help establish, later, meaningful relationships of group members. Echoing Jerram Brown (1975), Jane Goodall (1986) says there are three basic types of relationships: friendly (which includes caretaker and follower); unfriendly (which includes competitor); and sexual.

One can be friendly to achieve status or rank and, if that fails, there can be hostility. As Preston (2002) suggests, responding to another's emotion signifies that the perceiver understands that same feeling, so there is a sharing of somatic response and, importantly, mental representation. All of this is to say that prosocial responses require some moral learning.

Social creatures solve problems and conflicts similarly, from resource sharing and allocation to grooming and reconciliation. Simone Pika et al. (2019) claim that in the majority of wild male chimpanzees feeding on tortoises they observed, most involved "peaceful, cooperative food sharing." Primate groups, overall, make an effort for social cohesion with behaviors and norms that skirt closely to what humans might call moral conduct. However, that's an open, heated debate. Jessica Flack (2002) notes that what one calls morality is biological and has somehow evolved out of or alongside self-interest. She notes how famed biologist George C. Williams dubbed morality an *accident*. If this were so, why has evolution not altered it, especially how we see "moral" capabilities in great apes? Emphasizing natural feelings at another's suffering or situation, de Waal (1996a) finds sympathy and succoring in apes to family, clan, community, and even outsiders. Here's a speculative question, but one I've alluded to from the empirical data I've accumulated and the heart or *vera causa* of my book. In apes, could there be a connection between sympathetic behaviors and the mutualistic relationship they inherit and cultivate with their forest habitats? In other words, the ape land ethic can have emotionally "moral" overtones of extended altruism.

Citing the work of Peter Kropotkin (1902) on mutual aid and Robert Trivers (1971) on reciprocity, Flack (2002) notes that many animals engage in social behaviors verging on altruism. Cooperation pays off in the long run, such as sharing food, common among chimpanzees, bonobos, and orangutans. Animals grasp this fact. Sharing, then, raises questions about the apes' mental representations of the future, their planning, and the understanding of intentions, particularly in how they interact with the ecosystem itself. Even on the level of microbes, as Tim Hoek et al. (2016) note, accessibility to resources affects interactions ranging from mutualism to competition. In many instances, says Flack, apes don't simply tolerate theft of food but calculate past and future generosity among others. This behavior, in turn, raises what Trivers (1971) considers moral aggression, that is, a situation when not cooperating or violating sharing codes bears a negative cost. Aggression toward cheaters is witnessed by many and, therefore, helps establish favorable granting/receiving norms. The fact that there is enforcement of social stability signifies an elemental moral system accounting for individual and group interests. Obviously this social ethic among apes relates to their viable use of land in a bio-network, as I've been suggesting.

These sympathetic actions are based in empathic emotions and indicate how apes can make distinctions among individual identities and behavioral changes. It is not the mission here to discuss the evolution or biology of morality in full, since this has been covered (Tague, 2014 parts one and two; 2016, chapter 2). Suffice it to say that some thinkers, notably the esteemed developmental psychologist Jerome Kagan (Flack, 2002, Comments), question the conclusions of Flack and some ideas from primatologists like de Waal. In spite of what is shown here, drawing from the research of others, Kagan does not believe great apes hold good/bad concepts, feel guilt, or even express empathy. On the contrary, resource allocation and sharing for kin and others as moral behavior, it should by now be clear, are tied to how individuals and groups engineer the ecosystem. Many field observations (Goodall, 1971, 1986) and research papers (Brosnan, 2006) demonstrate, opposing Kagan, that apes are keenly aware of and richly react to inequities about fairness in the wild and in a research facility.

Brosnan was a student of Frans de Waal, and he goes on to say (2019) that elemental fairness is biological and not cultural. Many thinkers in this field would not discount the impact of culture, and I'm sure de Waal would not, either. His point seems to be that there is an "animal" linkage to what humans call moral behaviors. Hunter-gatherers also follow, says de Waal, a code of equitable food distribution evident among apes. Cooperation and fairness are not unrelated, and experiments on apes show this, where extreme selfishness is frowned upon. Apes not only monitor their own needs but are simultaneously sensitive to the desires of others, de Waal would say. Translate this mindset from the social to the ecological level and we see eco-engineering, where the forest is a gigantic food container an individual, among others, needs to rely upon without degrading. That is, the forest can be seen as part of the ape social ethic. On the other hand, many human societies see food resources as capitalistic enterprise, not sharing.

Of note, however, and somewhat beyond the "attitudinal reciprocity" of previous studies like Brosnan's, Valier Dufour et al. (2009) show how there is calculated reciprocity when favors are given or received, with increased complexity over time in tracking and weighing costs and benefits. Dufour's study deals with orangutans and reveals how exchange partners have and share intentions, comprehend the value of the items, and understand and agree to equitable payback so a balance is achieved. This exposes a heightened degree of prosocial behavior, meaning that exchanges and sharing are not essentially based in greed. There is a monkey, great ape, human continuum, but any moral sense, which is social, might have evolved differently because of group size, mobility, and the neurobiology of weighing memories of experiences in different communities, environments, and even according to gender. My aim is not to judge one type of society over another, whether ape or human, but the evidence at this point shows who the sustainable forester is and who is not.

Cheryl Knott et al. (2008) find female orangutan competition in Borneo in what are known as encounters of contest (overt aggression to gain advantage) and scramble (where shared food means equal reductions). Food competition among females is a selection pressure for reproductive success and infant nourishment. Mutual avoidance among hungry orangutans with overlapping ranges might even account, says Knott, for their semi-solitary nature. In fact, female primates are more dramatically affected by group living than males, note A. Catherine Markham and Laurence Gesquiere (2016). For example, there is group competition for food, which might necessitate more travel time. Status competition within groups dictates that those of lower rank have more costs and fewer advantages. Competition between groups can be damaging when resources are scarce. Stability and group size are affected when there's the imbalance of fission and the fragility of fusion. Whereas small group size has advantages in terms of less food competition and large group size ensures a probability of winning against outside competition, these dynamics certainly led to the increase in sentiments of approval/disapproval. If female primates feel the pressure of group size disproportionately, and since they rear offspring and nurture sociability, they exert incredible influence on an infant's reaction to the physical world and the behavior of others.

There is no so-called moral gene, but genes factor into what we call moral feelings. Rather, like humans apes use capacities of perception, deliberation, and decision to evaluate another's actions. For apes, not only did moral systems evolve in the forest, but they are still located there. Interactions between male and female, mother and infant, or kin involve tensions, resolutions, and mostly feelings of physical and psychological sensations. These are social sentiments and emotions which evolved into what we label moral behavior. Morality is a behavior, and behaviors evolved in ecosystems (Grant, 1963; Lorenz, 1965; Williams, 1966; Brown, 1975). In *Chimpanzee Politics* (1982), de Waal lays stress on how chimpanzees, especially males who are strong and always maneuvering for status, adhere to a "code of behavior" where dominance struggles in the unstable hierarchy yield very few serious fights. Today's foe is tomorrow's friend. More often there will be tantrums, screaming, and power displays. In any case, who wins any type of fight is less important than the coalition and support behind each battling opponent. Proving oneself to the group takes more time and effort than simply winning a confrontation.

A striking point is that moral senses (i.e., physical sensations and intellectual sentiments) are not some special ingredient added into "humanity." Similar to de Waal (2006), I see great ape emotional responses, behavioral decisions, and judgments as part of a long, evolutionary trajectory. This is not to say there is some teleology involved in reaching upward to humans. True evolutionary scientists dispute any notions of a hierarchical order that places humans on top. There is no top and no bottom, just an uneven plane of degreed differences, or descent with modification. Sociality, self-control, and maintaining order in and for groups in a habitat are very old adaptations and persist. As de Waal (2006) says, morality is not a human conflict with nature but is in nature. Indeed, de Waal insists that in discussing any natural moral senses we should place less stress on behavior, like food sharing, and look instead at the underlying emotions and intentions of, for instance, tolerance and reciprocity. I'd go further and say that some of the reciprocal sentiments underlying the ape ethic have much to do with how they correspond with their biologically networked environs.

While a fair amount of words throughout my book, and notably in chapter 2, hinge on great ape intelligence seen in communication and tool manipulation, the bottom line is that the paramount behavior of apes is how they deal with each other mostly on equal terms. For example, one chimpanzee understands when another loses a fight and offers consolation, a cognitive empathy where a highly self-aware subject can assume the perspective of another's physical or emotional suffering. How does this happen? While orangutans have feeding ranges that overlap, they try to respect boundaries. Why?

In answering questions like these, Sue Taylor Parker et al. (1994) say that great apes have a very advanced degree of self/other awareness. Robert Mitchell (1994) explains this in more detail. He asks how many dimensions of self anyone can have. The emphasis here has been on the character of moral individualism and the ape land ethic of ecosystem engineering, but let me take a look at what Mitchell says, mostly but not exclusively about humans. Mirror self-recognition should not be the gold standard of measuring self. For example, Mitchell details other aspects of self that include the following:

1. Perceiving self (e.g., vocals and the self-perceived, as in phantom limb)
2. Extended self (e.g., hair and gender)

3. Identified self (e.g., mimicking another's actions or mirror self-recognition)
4. Imagined self (e.g., planning, pretense, play)
5. Objectified self (e.g., intersubjectivity and emotions)
6. Presented self (e.g., comportment for influence)
7. Dissociated self (e.g., automatic behavior or self-abnegation when experiencing pain)
8. Evaluated self (e.g., autobiography)

As I argue, great apes, following this template (some of which could play into semiconscious ecosystem engineering), demonstrate self-awareness in accordance with the criteria of their group. These multiple selves manifest at different times on different occasions for different purposes, but all roll back to the stimulus of social interaction in a defined ecological environment. For great apes, let's add to the list of selfhood a bio-networking self by virtue of their intellectual and physical dexterity functioning in conjunction with the forest. Mitchell does not attribute all of these selves equally to apes, when he considers them, and even makes distinctions between signing and non-signing apes. It should be clear from the argument and claims in this book, though, that there is quite a bit of overlap between these human and ape selves. Previously, although I highlighted "languaged" apes to demonstrate their easy access to human culture, at the same time I've stressed how "language" is not necessary for the cognition, intelligence, social, or ecological proclivities of apes. The bottom line is that given these "selves" along with the capacities and abilities required to recognize and maintain them, great apes have earned forest personhood status, morally and psychologically.

EVOLUTIONARY ETHICS

What are the elements of social behavior relative to fitness or personal reproductive success, asks Robert Trivers (2002). One could include altruism, self-interest, cooperation, and even spitefulness. Reciprocity can be negative, requiting someone by doling out comeuppance. Fairness, says Trivers, is a regulatory consequence in selection for reciprocal altruism, that is, cheater detection, as opposed to fairness evolving from kin-kin exchanges. Whatever the case, could such social altruism and mutualistic exchange be extended to include great ape behavior toward their forest habitats? Contrary to some utilitarian thinking by Gary Varner (2012) following philosopher R. M. Hare, apes are not replaceable, and once gone there will be no megafauna to care for tropical forests as I've been arguing.

Why do humans imaginatively regard aliens from other planets as persons (e.g., Dr. Spok from *Star Trek*) but not great apes as persons? So asks Kristin

Andrews (2015), from whose chapter "Moral Minds" some of what follows in this section is derived. It's important here to expand on relevant ideas from my chapter 1. In terms of moral status, while the notion of personhood is important to him, Peter Singer (1975) emphasizes the suffering of animals more than their intellectual abilities. As a utilitarian, Singer would be more concerned about the larger good than individual rights. For Tom Regan (2004), in contrast, since an ape experiences a full life (or how he might say one understands the subjectivity of its own life), moral standing is deserved. As seen with Steven Wise (2000, 2002), having a moral right is critical to making a legal claim against someone else. Andrews explains how philosopher Gary Varner defines a person as one who is rational, self-conscious, autonomous, and considers outcomes and possibilities.

Varner (2012) thinks language to develop a self-narrative is central to personhood. This seems anthropocentric. In *Can Animals Be Persons?* Rowlands (2019), after John Locke, sees personhood in consciousness, rationality, self-awareness, and other awareness, all of which apes and other animals exhibit. Apes certainly grasp self and others in place and time. Varner ignores great apes like Washoe, Chantek, and Koko who learned to articulate their personal and self-constituting narratives through signs, symbols, and gestures, yet they did not have legal or moral standing as persons. However, as Laurent Dubreuil and Sue Savage-Rumbaugh (2019) show, enculturing mixed groups of chimpanzees and bonobos to respond to human speech and employ lexigrams to engage in dialogue is indeed, it seems, central to higher levels of consciousness. Yet, how much "language" would an ape need in its forest environment where she spends a good amount of time communing with trees, just as a mother tree seeds saplings with underground nutrients? As we know from cave paintings to children's picture books, a narrative can be established using only visual imagery or minor gestures stirring mental representations.

Andrews (2015) refers to philosopher Mark Rowlands (2012) who finds animals sensitive to their surroundings which elicit emotional responses. Rowlands believes an animal, not specifying great apes, can be a moral subject but not a moral agent with responsibility for acting. Since apes suffer and know harm, they are aware of responsible actions. Without citing all the academic or popular stories, though Frans de Waal's books are a good source, there are many instances of apes helping and caring for others and other species, even examples of risking their lives to save another ape or a human. In 1996 female gorilla Binti Jua, at the Brookfield Zoo, Chicago, rescued and consoled a boy who had fallen into her enclosure. Dale Peterson (2011) relates a story, via Frans de Waal, of female bonobo Kuni in an English zoo who found a stunned starling in her enclosure, climbed a tree to try and help it fly, and then, when that failed, guarded it from other bonobos until the bird was sufficiently recovered to take flight. Why do skeptical philosophers insist that ape emotions don't count since they are supposedly non-normative? Apes express norms of social behavior on a daily basis and demonstrate responsibility for their actions, and that is precisely why they have survived in linked subgroups, coherent groups, and large communities.

Rowlands (2012) says one must be aware in a moral state of achieving good consequences where the moral quality is driven by forces outside of the agent. When a chimpanzee patrol defends its territory by killing a hostile intruder, does not that act fill this requirement? There are, of course, more benign occurrences, as when a lone male orangutan or a silverback gorilla avoids physical confrontation. Do not those choices involve considering values peripheral to their own? We know that apes share emotional states with each other and can, even in captivity, experience emotional states with humans, proving they have moral sensitivity external to their own agency. Granted, perhaps, that an ape doesn't have a fully equipped moral dashboard with fancy concepts and terms as does a human being, but should she? Why are not many humans, by the same token, more ethically reliable and morally honest? Why are most humans not sincere ecosystem engineers? Besides, shocking to some as it might seem, the moral system apes employ, grounded in empathy, sympathy, kin selection, fairness, cooperation, and reciprocal altruism, is the basis of human morality.

Consider how in early human history among pastoralists and agriculturalists there was, according to bioethics philosopher Bernard Rollin (2009), a "caring" and vital "symbiotic" relationship to natural resources, especially to animals. Local environments had something akin to moral status. With the rise of large cities and chiefly industrialism, medical research, and factory farming most humans falsely justified morally relevant differences between themselves, the natural world, and especially animals. There are no morally relevant differences, says Rollin, to justify harming animals or people. Justifications include, for example, power over others, a privileged sense of reason over the intellectually or mentally challenged, or the claim that one group's interests trump another's.

All animals in their habitats have strong inclinations for their own welfare and even in the lives of kin relations. If some humans permit themselves to induce fear, anxiety, or pain in great apes by persecuting them or destroying their habitats for the sake of human economic desires, how is that morally justified? Science and industry are not, as Rollin (2009) would say, value free. Oil palm planters consider orangutans as "pests" and murder them by direct gunshot, and though that is outlawed in some places it still occurs (Lustgarten, 2018). Moral concerns are not only human. Apes have moral consciousness, as this chapter demonstrates, and avoid fear, anxiety, and pain for themselves and even for others. Great apes can question moral legitimacy. They know when to challenge authority, and they grasp the abuse of power or harms to the welfare of individuals or kin. Basic group behavior of primates is technically contractual. As I've been suggesting, all of this seems reducible to how apes hold a sense of conscience, as well, for their ecosystems.

There is an evolutionary ancestor connecting apes to hominins. Humans share with great apes the ability to cheat and deceive, as Frans de Waal (2006) would stress. Again, let me reference Andrews (2015) who sees Bekoff and Pierce (2009) emphasize outward behavior or what they refer to as "justice" where de Waal looks at the shared, underlying primate mechanisms. As shown in my book, while behavior is an important element, causes of the behavior need equal consideration. There is a difference, then, in the moral conscience toward land between humans and apes. Andrews goes on to note how for philosopher David Hume emotions are essential to and enable approval or disapproval of another's actions. Kant is more to rationality and acting on principles. Utilitarianism seeks to maximize good for many. All these views see moral agency differently, but they do get to fundamentals that Bekoff and Pierce sidestep by examining only particular behaviors of a species group. I tend to agree with Andrews who says that with an emphasis on behavior, Bekoff and Pierce, keen on species differences, don't allow for flexibility related to empathy, and in my argument ape sympathy for their habitats.

Akin to virtue ethicists, claims here are to the autonomy of an ape ethic of bio-networked moral individualism since it stems from the moral psychology of self-consciousness (metacognition) and perspective taking (theory of mind). As neuroscientists like Antonio Damasio (1999) point out, feelings and emotions, not necessarily rationality, are prime movers of moral behavior. Apes express and read from others a full range of emotions, whether for competition or cooperation. Andrews (2015) says theory of mind is not even needed for morality since social norms already exist in an ape society with intersubjectivity and perspective taking. On another note, in talking about morality, let's not refer to ideas from Plato but, following philosopher Lori Gruen according to Andrews, to simple daily acts like caring for oneself, an infant, or one's social and ecological habitat. These progressive attitudes of oversight are evident, as I've been showing, in terms of how apes approach ecosystems they cooperate with as if kin. At any rate, there are overly developed areas in Africa necessitating chimpanzees to cross busy roads, and as they do so at great risk, males have been seen to safeguard and guide others.

Among the scientists and thinkers mentioned in my book, an important philosopher, Martha Nussbaum (2006), relates how even in ancient times, 55 BC Rome, a crowd of enthusiastic people witnessing a staged battle between humans and elephants intervened and beseeched their leader to cease the wholesale slaughter feeling, in the words of Cicero, *societas* or fellowship with the pachyderms. I suppose this anecdote connects with what I've said about moral behavior in nonhuman animals by showing how humans can recognize, through empathy, their connections to us. This is in stark contrast to the stories Dale Peterson (2011) narrates about angry mobs in nineteenth- and early-twentieth-century America dutifully executing male elephants who in musth often killed a handler. Needless to say, why were the elephants taken from their habitats, confined, and put on public display to be paraded around for entertainment? Nussbaum refers to Article 21 of the Indian Constitution which requires a life of dignity, interpreted by some Indian courts, to include all creatures and animals. Reference is again made to the Northern Indian Court's (Sharma, 2018) decision precisely on this issue, cited in my chapter 1.

Nussbaum boils Kant down to something like this: Because of their supposed irrationality and, therefore, inability for ethical choices, animals matter only insofar as they are useful to most humans or how most people ascribe importance to them. In works like his *Metaphysics of Morals* (1785) Kant sees rational beings as persons and ends in themselves, principally because a human can represent himself with "I," thereby separating humans from animals. We now know, as I continually show, that apes think about themselves in relation to others, which qualifies as normative behavior. Furthermore, I am radically suggesting that apes also think about their interactive relations to their forest habitats, perhaps empathically. In Nussbaum's view, utilitarian philosophy is not much better than Kant's since it aggregates vast swaths of animal species and their fictional needs and desires. Nussbaum considers how there are capabilities of dignity in any sentient being, and the tragedy of stifling the potential of one's flourishing is of paramount ethical importance. That's an assessment made in chapter 1 of my book. Beyond the utilitarian concern with suffering and the Kantian focus on rationality, let me acknowledge the full range of consciousness, feelings, emotions, and daily activities any animal is entitled to through its complex eco-centered life history.

So Nussbaum is concerned about ethical justice for the livelihood of the individual being. Similarly, I raise this idea a number of times springing from philosopher James Rachels (1990) and his notion of moral individualism, critical to my argument. For Rachels, there's harm in reducing or eliminating a psychologically complicated being's autonomy, that is, one who is aware of the capability for freedom. Nature is neither good nor bad and should not be a prop for ethical theories. Rather, moving forward people need to build an ethics from the innate capabilities of a being, whether a human (e.g., a child or one whose intellectual capacities have been severely compromised) or an orangutan (e.g., free roaming in the wild or captive in a zoo). Perhaps it's too obvious to state, but worth considering. A good life for apes consists of autonomy and decisions in their own evolved, untouched habitats. The ape ethic of moral individualism is in the forest, not in a research lab. Orangutans don't simply want trees; they evolved to require a large tropical forest with all of its ecological strata.

How human ethical views influence orangutan reintroduction is a subject studied by Alexandra Palmer (2018) and related to thoughts in this section. Are human moral obligations for rescue only, which involves much time, energy, and cost? Should efforts be placed on fighting deforestation? As seen from the vignette at the beginning of chapter 1, ex-captive orangutan Judy simply did not want to be reintroduced to the forest after her rescue and rehabilitation. Palmer says that in anthropology there has been a move away from the study of social norms and toward how an individual conceives a more universal ethic of right/wrong conduct. This does not mean that one can offend moral norms while maintaining a rightful ethical responsibility. Rather, there is tension between what is explicitly moral and implicitly ethical, evidenced in ape social interactions and ecosystem engineering. There are evolved ethical justifications for humans to conserve resources and preserve natural surroundings, and so too with apes. My argument, of course, is that apes are superior conservation engineers of their bio-network than are humans. Orangutans, for example, don't need a political or economic reason to maintain a stable ecology, but, on the other hand, many powerful humans demand at least those reasons. Although there is a shared biology of morality across humans and apes, the ape land ethic on the evidence cited throughout this book appears far less menacing to the environment than the human ethos of rampant development, mass production, and runaway consumption and waste.

This brings me to another relevant question in the moral realm, which I ask at the beginning of the next section.

SOCIAL CONFORMITY

Did brains and cognition evolve because of group pressures for intense social competition and complex social relations? Because the human brain is so large (2% body weight and 20% energy usage) it cannot have evolved, as with other animals, simply to survey and process information about the environment, although that is surely a critical factor. Since great ape brains are large and neurologically complex, there are evolutionary links between them and the human brain. Deception and coalition forming, says Robin Dunbar (1998), run across primate species, and the social brain hypothesis suggests that consciousness and reasoning were selectively improved to include an ability to make social calculations, as with theory of mind. Like humans, great apes operate along scripts of behavior and knowledge about event sequences, for example, when one goes to A, y happens, says Robert Mitchell (1999). Scripts are representations from social activities or interactions one has experienced often, as in routine movements in the forest. So, one can, therefore, use a script to connect the dots, meaning one can create a new script for Ay if its pattern resembles Av. Knowing scripts, says Mitchell, allows one to enact deceptions, where the benefit falls to the deceiver.

"Machiavellian intelligence," says Frans de Waal (2016b), is too narrow a term and excludes social knowledge that encompasses cooperation and empathy. Contradicting Tomasello (2014) from chapter 2, de Waal says that studies beginning in the late 1930s demonstrate cooperative behavior where one chimpanzee who knows another helps for little advantage, suggesting anticipation of future gain via direct or indirect reciprocity. (See, for example, Trivers, 1971; Alexander, 1987.) In my argument, as expected, I'm saying such ancillary receipt occurs between apes and how they "cooperate" with their forest habitats. In tests where multiple chimpanzees, whether an established or new group, are involved in pulling a rope for a common goal they usually succeed, discouraging cheaters and freeloaders. As de Waal says, in order to cooperate, chimpanzees will manage conflict. Indeed, in a natural environment where there is minimal separation with few barriers, where the effects of one's actions are clear, where there is reduced reward competition, where all choices of advantages are visible, and where there is a high degree of communication among subjects, it has been verified that chimpanzees, often viewed as the ultimate competitors, will act in a prosocial and not in a selfish manner (Horner et al., 2011).

In other experiments, chimpanzees cooperate to share limited resources, behavior not limited to one ape species. Following from this, we can say that apes are not impulse-driven but are self-conscious, self-aware, and exercise restraint and control in planning. There is some qualification here, though. As Anne Russon (2004a) points out, orangutans are far less impetuous and rash compared with chimpanzees. The inhibition of aggressive behavior, especially, is not just social but the basis of moral individualism. Apes are not types but, like humans, distinct persons. It is well known that there are wide individual differences among apes, as explained, for example, by de Waal (1982) and Goodall (1986). These differences range from faces, voices, and traits like walking, sitting, or other bodily movements. This is to say nothing about temperament and intelligence, or mood and attitude, which explain the complexities of the social group and the need to keep track of individuals. Apes indeed are persons in how they manage each other's biography.

Culture, or variations of behaviors across genetically or ecologically similar communities, is no different than other biological functions, like grooming or communication. In this regard Victoria Horner et al. (2009) are critical of studies that use human infants guided by human proctors while the child's parents are present. There is usually no comparable arrangement with tests on apes. There's a selective advantage among apes, especially chimpanzees, for conformity behavior, and this gets lost in lab observations of captives. In this way, culturally, learning can lead to group behavioral differences. One conforms not to be equivalent to others but for strategic reasons regarding social survival.

Catherine Hobaiter et al. (2014) show evidence for the social transmission of tool use in wild chimpanzees. Learning is dynamic and occurs over multiple observations of a novel behavior. This was seen among Sonso chimpanzees, Budongo Forest, Uganda, who culturally shared moss sponging and leaf sponge reuse, learned socially and spread through a network. In one field study (Luncz, 2018) it was determined that a migrating chimpanzee will pretend to be ignorant about how to crack nuts simply to fit in socially. She knows from her prior group the optimal method of nut cracking, whether to use a stone or wood hammer or whether to use precision taps as opposed to hard strikes so as not to smash the meat of the nut. However, the desire for group inclusion will be an incentive to employ a more costly and less effective method. Like humans, some apes have such a motive for social inclusion that they will bow to peer pressure, as Luncz shows. All along I've been suggesting how apes behave inclusively with the forest community itself. Neighboring chimpanzee communities of the same population, say Kathelijne Koops et al. (2015b), exhibit differences in tools for dipping. Is this cultural or environmental in genetically related chimpanzees? The selective tool use apparently has nothing to do with the similar ecology between communities and so is cultural. Anticipating Luncz, Koops notes how chimpanzees have been observed to change their tool use behavior when migrating to a new community. This social conformity, I say, extends to ape behavior with their habitats.

Though humans share recent evolutionary history and genes with chimpanzees and bonobos, there's always a mix-and-match descent. Koops et al. (2015b) examine the question of motivation for tool use between bonobos and chimpanzees. Bonobos use fewer tools and don't usually forage. Ecological and social opportunities do not explain the difference, since Koops says it's doubtful that bonobos have a ready and accessible supply of food over chimpanzees. However, chimpanzees and bonobos inhabit separate geographic areas which might explain tool use differences. It seems chimpanzees from an early age are predisposed to play more often with a variety of objects. Variation of tool use across chimpanzee populations would then be explained by culture. Bonobos typically feed together, especially when young, so they might be more tolerant of others. Whereas bonobos are typically more in tune with social cues, chimpanzees tend to focus on objects and goals. Orangutans appear to be motivated by human-based signals as opposed to artificial markers, meaning that they readily respond to social cues (Marsh, 2012). The upshot, certainly, is that any ape "culture" is immersed in their reciprocal exchange with the forest.

There might also be evidence of ritual social behavior in chimpanzees. Beyond the disputable rain dance of a chimpanzee group (Goodall, 1971), there is, more recently documented by Hjalmar Kühl et al. (2016), stone accumulation and stone throwing. This is another instance of primate archaeology, mentioned in chapter 2 regarding 4kya chimpanzee stone technology. Here, the accumulation of stones is reminiscent of the construction of cairns by humans. In West Africa, the researchers discovered, chimpanzees routinely behave in a ritualistic fashion as they bang or throw rocks at trees and into tree cavities, where, once fallen, the stones are then deliberately accumulated. Stone throwing, not confined to chimpanzees, is considered agonistic display in captive Japanese macaques, wild baboons, and female capuchins. Chimpanzees, however, exhibit the greatest variety of tool use. Considering the wear on the trees and stones, this throwing behavior has occurred over long periods of time, and the same chimpanzee would repeat the movements at the same site. As part of the ritual, there is pant hooting and drumming, followed by waiting and listening. Since the behavior is spotty, it's probably not genetic but cultural. If this behavior is not completely for agonistic display, then there is some symbolic communication involved, grounded in emotional and intellectual intensity perhaps driven by social-ecological correspondence. Ape culture is rooted in their land ethic.

Demonstrating their moral individualism, great apes are not without very strong emotions. Like humans, a bonobo's attention gravitates toward socially emotional images and scenes. Mariska Kret et al. (2016) say bonobos focus on others who yawn, groom, or are engaged in sexual activity. Unlike humans, the bonobos' attention does not seem distracted by aggression or distress. The explanation for this difference might be that bonobo social life is heavily geared to affiliation. In contrast to humans and chimpanzees, bonobos can better regulate their emotions to avoid forceful conflict. Demonstrating species continuities, all great apes, says Kret, are capable of identifying human emotions. Beyond identification, let's say that there can be a reconciliation of emotions across human and ape species, demonstrated by field researchers who have engaged chimpanzees, gorillas, and orangutans in interspecies communication. Anyway, while chimpanzees are clearly sensitive to the emotions of their peers, bonobos are more tolerant, better at eye contact, and much more risk-averse. Gorillas, on the other hand, tend to avoid eye contact, seeing it as aggressive. Kret also shows, compared with the chimpanzee brain, how bonobos have more amygdala (reactivity) and insula grey matter and so perceive emotions better. Similar to the human brain, bonobos have greater flexibility between the amygdala and anterior cingulate cortex (self-control) and more Von Economo neurons for social cognition.

It's hard to tell, from some of these studies I cite, if apes act like humans or humans act like apes in terms of social behaviors, and hence the question of ape personhood. Ultimately, all ape knowledge (inherent and learned) and communal behavior is grounded in ecological activity, and that's the main difference, where apes are tenable stewards of their habitats.

Integral to consciousness is self-awareness with psychological and behavioral consequences, where the sense of self is primitive from an evolutionary and comparative perspective. Louis Moses (1994) says that recognizing one's self means subjectively feeling and objectively thinking. Mirror self-recognition is what it says, visual. While important as a gauge, it is not the only means by which one recognizes a self. Awareness of a mental self in space in the choreographic movement of other minds and bodies is equally important and lends credence to the idea of an ape environmental ethic as moral individualism. There is an epigenetic component, or modification of gene expression and not gene code, Moses suggests, to the notion of self as part of a community of others. For instance, in 1,000 test trials with captive, young chimpanzees, E. W. Menzel (1971) learned that these apes easily communicate specific information to others regarding the whereabouts, physical status, and general value of objects. There is, suggests Menzel, a "language" in this communicated attention where meaning is made in the mind of whoever receives the message. Signals can be expressive (i.e., emotional) or referential (i.e., informative). Signals are intentional and carry thinking about another's beliefs. Of course, I'd extend this social "language" of apes to say it is part of the vocabulary they use to communicate with forest bio-networks.

Following another's gaze is social orientation or indirect contact with others and is probably evolutionarily old. From infancy, rhesus macaques exhibit this tendency, say Alexandra Rosati et al. (2016), but the behavior increases in juveniles, is shared mostly among females after sexual maturity, and declines along with age. This developmental pattern is found in humans, too. Gaze can be characterized as a form of social-cognitive, deliberate information gathering about others and community situations processed and stored for later reference. Imagine, too, how this gazing behavior is employed to feel the mood and atmosphere of the forest trees and vegetation. From gaze behavior in other primates more complex theory of mind likely evolved. Gaze awareness and the other types of conformity outlined in this section are vital signs of moral individuals competing and cooperating for resources on the same turf.

Social living has benefits of protection and shared resources. In this way, says David Buss (2008), apes have formed alliances to cooperate, giving rise to reciprocal altruism. Altruists outcompete the selfish, and so the trait spreads. We see this in forms of group foraging. Outside of kinship, one who gives help is most likely, later, the one to receive aid when needed. Could there be a form of altruism apes share with their green biological networks, as I'm proposing? With male chimpanzees a reciprocal alliance with a female is necessary for status. In addition to cooperation and reciprocity, there can be indirect or delayed reciprocity. Apes are capable of identifying individual group members and especially recognizing another's values. Cheaters are detected and often punished. In groups, it's important to have and model shared values.

Having considered evolved moral structures, I want now to focus on the social personhood of apes in terms of their motivations, traditions, and learning related to forest ecosystems.

GREAT APES AS SOCIAL PERSONS

Most social animals, including all great apes, are actively obliged to conform in any number of ways, whether through parental-offspring units, coalitions, small groups, or the larger community. Do differences of behavior between humans and most animals justify asserting the moral relevancy of one species over another? So asks Bernard Rollin (2006). As with humans, a common denominator for many mammal species is sociality. There are differences among individuals and across human societies, yet the majority of people tend to view nearly all humans as entitled to basic rights because of shared norms. Consider, now, the close behavioral and psychological similarities between great apes and humans. Why do apes not have extended rights? Rollin and others, for instance, Peter Singer (1975) and Tom Regan (2004), assert that animals deserve moral consideration. Objectors would say that animals, especially since they are without "language," are not part of moral contracts. More bluntly, the objection goes, animals are not moral agents and only recipients of moral consideration bestowed by their human superiors. Laurent Dubreuil and Sue Savage-Rumbaugh (2019) help show, in fact, how apes are "languaged" moral individuals in their sustained, long-term, group and bicultural communications. Moreover, Eva Meijer (2019) shows how animals routinely communicate with humans in ways demonstrating their rights to democracy and citizenship.

From its etymology, the word "social" includes elements of alliance and friendship. Beyond the chief notion of ape-human continuity, which I have been rehearsing throughout this book, I want to ask readers to consider, in line with my developing argument about ape-human difference, how apes act with and react to their habitats in a mutually inclusive social manner. It's as if they are aware, on some level, how their genetic heritage and longevity are allied to the continued health of their forest homes.

Agency, of course, is not the sole determinant of moral consideration. Those humans mentally impaired, those in a comatose state, and those with severe dementia, as well as infants, are not really conforming moral agents and yet are considered persons. Like humans, animals have interests related to their needs and avoid suffering. There's no requirement for animal "language" per se, especially not in line with a so-called social contract. Animals have evolved and developed group rules they follow; as one example, there's the mutual maintaining, obtaining, and sharing of food resources in a range. Rollin (2006) points out that animals exhibit moral agency when one relents after attacking if the other shows submission. This is true, essentially, with great apes and humans.

For Rollin (2006), the crux of the argument once again comes down to how Kant chauvinistically insists that moral agency and moral concern both stem from rationality. Moral agents must be rational, so the claim goes, and only rational beings can be the subject of moral concern. Some key social decisions are emotional, not rational. Based on the information accumulated and discussed in my book, it should be evident that great apes, indeed many other "animals," do not operate solely on a stimulus-response system but deliberate, make choices, and then reflect on that decision in similar situations later. Reasoning is not tied to stimuli but is associated with concepts to help the ape conform and avoid conflict. An ape decides how to move or behave, at times, because of mental maps or social norms.

In other words, individuals and not just species are at the center of moral concern. An ape's self-fulfillment is grounded in the freedom and willingness to move and act in her habitat and among others. Apes have physical, emotional, paternal, and sexual interests, or what consciously matters to them. They are "morally" concerned to propagate their interests, for example, of ranging for resources, communicating with others, and avoiding pain. As Rollin (2006) says, these viable interests are akin to one's *telos* or nature. Great apes have free will and moral individualism in the awareness of their own intelligent choices and interpersonal interactions. Natural, moral law is dictated by the collective values, social behavior, and ethical norms of a community, not imposed upon them. This is not a utilitarian philosophy but flows from the *telos* and rights of the individual as recognized by others. Even when one ape behaves aggressively toward another that is an acknowledgment of rights, for such an act is a conscious infringement on someone else's welfare.

The question of personality in its full sense means a being with emotions, mind, and will. Does a stone have personality, or a worm, or an automobile? Personality, not persona, is integral to a natural person, an animated entity who can possess rights and transfer duties with an understanding of them, a lawful actor entitled to confer benefits and capable of liability. An ape's personality is shared with and tied to its wooded bio-network. Jane Nosworthy (1998), for one, argues that society would advance in social acceptance and, therefore, benefit as a whole if it were open to conferring legal personhood to great apes. Then, Christine Korsgaard (2013) reminds us that Kant believes only humans, because of their rational thought, choose how to live.

This Kantian way of thinking ignores the routine paternal, ambulatory, dietary, sexual, social, and survival choices of great apes. These choices are grounded less in instincts and more in learning and intelligence. In human life, as evolutionary psychologist David Buss (2008) shows, many short- and long-term mating strategies, perceived to be choices, are driven by evolutionary psychology. Of course this does not exclude choice, as great apes make mating choices too. So humans are biologically motivated to pursue self-interests yet self-deceptively pawn off evolutionary geared choices as wholly autonomous. Where is Kant's rational autonomy there? Korsgaard (2013) suggests that the conundrum of ape personhood is in how legal definitions have been established and set precedents over hundreds of generations. Meantime there are apes like Chantek, Koko, Kanzi, and Washoe who behave like good human people. In addition to the legal concept of person consider an ethical concept of the moral individual, a being recognized as inherently worthy of respect and dignity.

There is an evolutionary basis for personality and what in psychology is called the five-factor model of traits ranging from openness, conscientiousness, extraversion, agreeableness, and neuroticism (including their implied opposites). A study by Robert Latzman et al. (2015) examines the grey matter volume and asymmetry in the chimpanzee cortex which provides an empirical basis for confirmation of neurobiological correlates of personality. In an earlier study (Latzman et al., 2014), in addition to the five factors, a sixth of dominance could be added to chimpanzees. Alexander Weiss and James King (2015) confirm the five-factor personality model in great apes, with an emphasis on the differences between males and females evidenced more strongly in chimpanzees and humans than in orangutans. Like humans, extraversion and neuroticism decline with age in chimpanzees and orangutans, though there is a decline in agreeableness in orangutans where it increases in chimpanzees and humans. This is not surprising given the orangutan's semi-solitary nature, perhaps an adaptation for foraging success, and the chimpanzee and human political nature. In turn, male chimpanzees display higher neuroticism than male orangutans, also not surprising given the data about interpersonal and group interactions I've presented. While there is homology of personality development in hominids, it has evolved differently across species in accordance with their social strategies in turn affected by their physical habitats. The ape personality is one of inclusive social communication with the ecosystem.

In line with studies applying theories of Piaget to apes (Parker & Gibson, 1990), the fact that psychologists can readily assign temperamental types and personality development to great apes reveals much about their personhood. As a cautionary note, however, readers should not think of orangutans as misanthropes. They are not. Barbara Harrisson (1962) relates numerous examples of how her young, male ex-captive orangutans preferred each other's company, to the point that they would deliberately sleep together. Orangutans, say Gisela Kaplan and Lesley Rogers (2000), form four types of social groups:

1. Mother/infant;
2. Consortship between male and female, which could last from a few days to a few weeks;
3. Travelers in small groups;
4. Temporary group formation, as when they feed together in trees.

Orangutans tend to be more solitary in the higher regions of Borneo, where most tend not to populate anyway, than in Sumatra where more rich foods will attract groups in trees. While it is true that chimpanzees and gorillas will commit themselves to bonds or groups, this does not mean that orangutans are not social animals. Rather, they have adapted to their treetop environment differently and more constructively than other apes, including humans.

In any case, Latzman (2014) notes that recent work sees only four personality factors in chimpanzees: dominance, extraversion, conscientiousness, and agreeableness. Key here is the parallel between human/chimpanzee personality; though see, too, Tan (2017) who says bonobos can act sociably to those from another group. Overriding the five-factor model in a hierarchical relation, at any rate, would be three basic temperamental traits: negative emotionality; positive emotionality; disinhibition (vs. constraint). For example, says Latzman (2014), neuroticism and agreeableness create negative emotionality; extraversion and openness make positive emotionality; low agreeableness and equally low conscientiousness create disinhibition. The point is that these personalities are evident in humans as well as apes and regulated, in some part, genetically via vasopressin, a sex-specific neuropeptide across mammals implicated in aggression and territoriality (especially males), pair bonding, and altruism.

In humans, some brain regions are important to personality, basically the frontal cortex and particularly the anterior cingulate cortex and the prefrontal cortex. With the anterior cingulate cortex, behavior and motivation are cognitively regulated. With the prefrontal cortex, action appears oriented toward goals and is, therefore, related to other brain regions of attention, motivation, planning, inhibition, and memory. Latzman (2014) concludes that there are neuroanatomical correlates between apes and humans. The primate brain is adaptable, and is, according to Pierre Enel et al. (2016), able to reservoir neural connections in the prefrontal cortex to dynamically anticipate behavior representationally. Brain regions evolved in an environment of social adaptation, gestural and mental communication, ecosystem engineering, and overall strategizing.

Is an ape's inner life like a human's? We don't know for sure, but how an ape agrees with and is open to cooperating with her environs is a key difference. This is why I call for apes as persons of the forest who, utilizing their advanced "human" capacities, have diverged to become eco-engineers par excellence.

We can know how an ape behaves but can only guess at how she really feels, unless she has been taught sign language. Work like that of Fouts (with Washoe the chimpanzee), Patterson (with Koko the gorilla), Miles (with Chantek the orangutan), Savage-Rumbaugh (with Kanzi the bonobo), for instance, offers a window into the great ape mind. Like prelinguistic children, apes organize their world through nonverbal motor behaviors and react on social cues from others. There are many layers of social contexts (e.g., empathy), so it is not just observable behavior that counts but the mind and feelings behind the behavior. Nonetheless, consider a study by Claudia Rudolf von Rohr et al. (2015) that shows how female and male chimpanzees react strongly, by screaming or intervention, when seeing acts of harm against infant chimpanzees, demonstrating social norms. Feeling for another's welfare is not, therefore, only a human capacity and points to the biological roots of what philosophers call morality. The evolution of altruism is no doubt connected to the evolution of sadness displays and shared sympathetic emotions in great apes (Preuschoft, 2000). My guess is that apes, considering the environment of their evolutionary adaptation, express empathic moods toward their verdant habitats. Humans, too, can respond emotionally to nature, but as is evident from the climate crisis, historically they've been poor ecosystem engineers.

In many of his books, de-emphasizing the notion of a "selfish gene," Frans de Waal (e.g., 2013) relates stories of how chimpanzees and bonobos express benevolent care and even compassion for friends or kin in need or who are gravely ill. As already alluded to but worth emphasizing, chimpanzees have been known to care for orphans, a motive with no ostensible benefit. This altruistic, prosocial behavior has been documented in fieldwork on wild chimpanzees by Christophe Boesch et al. (2010) for over thirty years where at least ten males adopted motherless babies. De Waal says, drawing from James Rilling of Emory University, that neuroimaging experiments show a bias for emotional cooperation that must be overridden by strong cognitive efforts. In other words, altruism is a spontaneous, natural impulse that makes one feel good, even counting ape oxytocin, and need not be costly or calculated for selfish gain. Would that not count for personhood?

Since there is some overlap between mental and social behaviors, one point of this chapter that also highlights the important subject of ape moral individualism would be an examination of the so-called philosopher's brief, filed as *amicus curiae*, in the New York State Court of Appeals (Andrews et al., 2019; Nonhuman Rights Project, 2018). Courts weakly and arbitrarily use criteria of species membership, social contract, community membership, and mental and social capacities. By now readers should be able to discern that great apes exhibit mental and moral behaviors across these categories that squarely place them in the arena of persons. Some might claim apes are not part of a social contract. For significant social contract thinkers, like Thomas Hobbes, John Locke, and Jean-Jacques Rousseau, the social contract is intended to restrict freedoms. Social contract theory requires persons to bind to the contract. Contrary to Hobbes, life is not necessarily nasty, brutish, and short without any "social contract," for how then does one explain the propagation of many species over countless generations? Even in the wild, great apes live for decades peaceably among each other. Social contracts in themselves are not the source of rights and liberty, for those are inborn and usually compromised by entering into the artificiality of adhering to rigid group norms that conflict with peaceable individuality. Besides, when is rationality or social contract theory the arbitrary determination (and faulty at that) of personhood? Why not survival and reproduction over time in a mutually cooperative ecologically shared community? Ants and bees, following this line of thinking, would be persons, but the point I've been stressing is that since great apes operate as eco-engineers on such a large scale with positive, wide ranging environmental impacts they should be granted dominion over tropical forests.

Instead of talking about apes, we should listen to them, as do Laurent Dubreuil and Sue Savage-Rumbaugh (2019) who dialogue with apes on multiple levels to perceive their motives, needs, and desires. Where some have done extensive fieldwork, it is shown that apes, first, establish their own social/ecological communities and, second, have been limited by some human encroachment. It is now vital to ensure that these communities of contiguous lands become designated as separate spheres, since cohabitation, up until now for the most part in many places, has injured forests while exclusively benefiting some human groups. In another respect, incarceration has not obviated great ape personhood. Like people, they are social creatures, autonomous and rational, key elements of community membership and personhood. Autonomy includes rational components like intentionality, self/other awareness, and inference abilities, all of which apes possess abundantly. Autonomy is connected to ethics. In most cases, to eliminate one's autonomy is unethical. With autonomy one acts ethically or not. An ape ethic consists of, for instance, moral individualism, parental care, bonding and grooming, group activities or engagement with others, territorial patrol, and resource maintenance and sharing. Considering the mass of data from scientists and the informed opinions of philosophers, human societies will soon have to buckle to the obvious and grant ape personhood since it is evident as pictured here and elsewhere.

To win the court battle for great ape personhood, as suggested by Shawn Thompson (2019a), scientists and philosophers need to focus less on ape-human biological continuities and more on public policy. At the same time, this is an issue of public policy since, in spite of clear differences, great apes, according to Jane Desmond (2012) easily replicate what one designates as human. As I aver, how ape persons and their forest homes are treated should be a matter of concern for all people with a social conscience. At the least, as primatologist Christine Webb, ethicist Peter Woodford, and animal behaviorist Elise Huchard argue (2019), the divide between practicing behavioral science and considering animal ethics should be narrowed. To rephrase them, behavioral scientists who study the sociality of animals need to consider ethics and engage in conversation with philosophers. In spite of mounting evidence about a spectrum of animal feelings, consciousness, and moral behaviors, many ethologists and animal psychologists have been passively involved in the debate about animal ethics and have permitted nonscientists to innovate this field.

For instance, as my book elucidates, ape autonomy fits into cultural anthropology because of capacities and mechanisms like rational thought, self-awareness, and reciprocity. The threshold for personhood is low and is met, especially in how apes act with intention in an informed manner without undue influence, as often related here. This means apes act autonomously even in situations of control or in status disputes. Ape intelligence, then, mitigates their emotion in a control situation since they are not robots. This is especially true among orangutans who decide to socialize or not, which could be interpreted as emotional intelligence. Great apes, as discussed in these pages, have innovation, altruism, and self-control, hallmarks of autonomy in a social community. Since apes are "housed" in zoos and labs, the latter for psychological or physical experiments, it is in the interest of scientific public policy to recognize them as persons and not as prisoners. The benefit here is that, as I have been claiming, if apes are given forest personhood to maintain their domains, the biodiversity in those ecosystems will thrive; such flourishing is advantageous not only to the forests and beings who live there but also to how the forests provide a necessary cleansing of atmospheric carbon which in turn effectively regulates air and ocean temperatures.

Likewise, Anthony Rose (1995) suggests that primatologists need to influence public opinion to leverage decisions about keeping ape habitats safe. This involves not simply describing an orangutan in her environment but in showing how she reacts with conscious awareness and prosocial tendencies in her ecosystem, as I'm trying to do in my book. Reporting on the environmental ethics of Christian humanists, Daniel Cowdin (2008) relates how these types of scholars say relationships in a community involve participatory, reciprocal actions of mutual sharing and learning regarding family and others. These apologists, however, offer a veiled anthropocentrism of the Western tradition and assert that such communal values are necessary capacities for human persons to flourish. The view represented here, on the other hand, asserts that the notion of the flourishing person can be equally applied to great apes, since they exhibit a "public life" in their explicit moral individualism as social eco-engineers.

SOCIAL MOTIVATIONS AND UNDERSTANDING

Pulling from the previous chapter and the earlier sections of this chapter, let me now address how cognition, intelligence, and social behaviors are not only related to each other but are also connected to the ape ethic of moral and ecological individuals.

Darwin recognizes and understands the expression of cross-species emotions. In chapter 2 of his book *The Expression of the Emotions in Man and Animals* Darwin (1872) shows how dogs and cats express rage, anger, sadness, and happiness. An angry dog communicates his emotion to a cat, which might not particularly care though she comprehends the gesture, notes Darwin. Monkeys, too, exhibit an array of facial and bodily expressions and postures that reveal and communicate emotions. As Signe Preuschoft (2000) notes, there is a "correspondence" of understanding between signaler and signaled. Is it any surprise, then, that great apes who are genetically close to humans would be motivated to express their emotions with similar signals? Cross culturally among people there is a range of equivalent emotional expression and body language, proving they are innate and inherited. In this way emotions are, whether for humans or apes, social signals to rouse, join, challenge, or share feelings. While feelings and emotions can come spontaneously, there could be mental representations as afterthoughts for future reference. Social motivation is multilayered across human and ape persons. Feelings can be moderated by other feelings or by the intellect, for altruistic or self-interested reasons, in both humans and apes.

In order to make sense of their complex social worlds, even among the semi-solitary orangutan, mental processes are in high gear. Apes interact on a social level and, no doubt, imagine such interactions, which in turn affects future behavior. Impressions are formed where an individual will avoid another or react aggressively/affectionately in his or her presence. In this way, there is an *affective* response that can, or not, be controlled. Social representations are evaluated and modified as needed on a continual basis with any species that interacts with others in or among groups. What I've been offering all along is that apes might mentally interact with their forest habitats as well. Primatologist Leif Cocks (2016) tells the story of a female orangutan who gave birth at the zoo where he worked. She let him hold her infant because he had regularly spent time in her enclosure. This was not a rare occurrence.

With other births Cocks was permitted by the mother to hold a baby, as well as letting the infants grab his fingers. In addition to whatever social emotions or personal feelings this orangutan person had for Leif Cocks, she clearly held positive images of him and his potential actions from prior experience and was so motivated to act complacently.

In addition to representations about any individual or event, an ape may take that knowledge and apply it generally. There are mental patterns formed across brain areas, for example, memory, fear, executive thinking, which can then be scrambled to establish meaning based on situational activation. Some of these highly emotional social evaluations, therefore, happen outside of awareness and can be harder to control. Take, for instance, a strange ape apparently from another group who approaches an immature male. There might be an automatic response of resistance from the youngster. At the same time, motivations and incentives can have dramatic effects on judgments. If the stranger approaching holds much desired fruit, and if he's alone, perhaps approach rather than withdrawal is more desirable.

Depending on the outcome of this situation, future social motivations will be cognitively altered. Strongly held emotions and memory affect cognitive motivations. Like humans, ape social behavior depends not only on the representation (real and imagined) of the other, but also on one's own self-image and status. If the confrontation with the stranger becomes aggressive, does the young male have a coalition of allies nearby to assist him? If he's low in the hierarchy and miscalculates an attack on someone well-liked and highly placed, how will that affect his future behaviors? This type of mental-social calculation in the primate mind is not far-fetched, for one only has to read field narratives of Goodall (on chimpanzees), Fossey (on mountain gorillas), Galdikas (on Bornean orangutans), or Strum (on baboons) to find such individualized examples. Categorizing individuals is a cognitive behavior which can have dramatic results not only on an individual but also on the group. Biruté Galdikas and Gary Shapiro (1996) tell how a profound interspecies event creates a mirror effect, especially in the way in which the feelings and emotions of an orangutan and human are expressed and mutually recognized.

A profound interspecies event might even work both ways, evidence of distinct yet shared social personality. I'd go as far as conjecturing that what we call a profound interspecies event happens with apes as they relate to their animated forest bio-networked environs on physical and through psychological dimensions.

Darwin (1871, chapter 3) correctly demonstrates that animals share many of our senses and emotions, and he lists, for example, jealousy, deceit, suspicion, emulation, wonder, curiosity, gratitude, and magnanimity. These are evolved behaviors that have adaptive value and are not only human. I'd go further and say, as I've been elucidating, that differences are crucial, notably in how great apes rationally and emotionally express themselves toward their tropical habitats. Darwin explains (1871, chapter 4) that some animals, as if reasoning, hesitate in order to consider their emotions, engage faculties, and weigh alternative actions. In their research, Richard Byrne et al. (2017) demonstrate how great ape gestures are intentional communication of feelings. Consider how across many mammal species, and particularly in great apes, facial expressions operate to communicate emotions and thoughts, much as in humans. Facial expressions, too, are evolved social adaptations. Going further, Mariska Kret and Masaki Tomonaga (2016) claim that chimpanzees visually process butts in a configurable manner like faces. Darwin also justly indicates (1871, chapter 4) that human social instincts predominate over those which are less persistent.

In other words, since apes, including humans, interact socially and mentally quite often, *mores* regarding engagement have become more evolved than predation tactics. Could there also be, I speculate, ape moral customs related to their biomes (in distinction to human)? Many other mammal and especially primate species socially operate in similar ways. There is motivation for care and concern about how one moves in a group or even how one wants to seem to appear in a group. Where people have a moral sense, approval/disapproval responses appear in other species to the extent that great apes will engage, at different times, in moral aggression or self-control. Social instincts in animals are evident from how they enjoy each other's company and will help each other. Importantly, there is a very high degree of nurturing from mother to offspring, most likely from which arose forms of altruism and human moral sensations (Hrdy, 2001, 2006). My estimate is that for apes this prosocial modus operandi finds equivalency in their interactions with arboreal tracts of land.

The question of the evolution of self/social-control is important and worth discussing (Boehm, 2000). Human moral systems help individuals in groups. Alexandra Rosati (2017) shows that all great apes are able to exert self-control for rewards so as to inhibit gratification. Audrey Parrish et al. (2014) demonstrate that orangutans delay gratification even when food is visible. This is significant because it demonstrates, in contrast to some lesser primates but in comparison to human children, that apes employ cognitive motivational controls for strategic, future-directed decisions where rewards cumulate. This is especially true in chimpanzees who, in social contexts, can be quite patient in waiting for rewards. How has the primate brain evolved and what were the selection pressures that helped increase the cortex and executive functions? Like humans, great apes have, in addition to the so-called limbic or mammalian brain that coordinates important bodily functions, a large frontal cortex that evolved because of selection pressures dealing with motivations for cooperation and competition. For the purposes of this book, translate that thought to ape collaboration with their habitats.

A long-standing theory which has personhood merit but does not accord fully with ecosystem engineering is the social brain hypothesis. This means that the cognitive demands of recognizing, establishing, and sustaining social relationships enhanced brain growth. However, one would be negligent to exclude other hypotheses that focus on factors like foraging and overcoming problems of resource recovery in an ecological niche that also impact on brain growth. In fact, Lauren Powell et al. (2017) go as far as saying that range size, and not sociality or diet, has been the main driver of brain expansion. Keep in mind, though, that ranging and foraging can be, broadly speaking, social as well, not only with conspecifics but with elements of the land. One needs to weigh the costs and benefits of feeding in another's area or with another group. As Hannah Mosley and Michael Haslam (2016, p. 70) say, the key outcome for cognition is that "something . . . changes the physical world as it happens." Though these hypotheses might not land squarely on self-control, they nonetheless deal with the expression of control, whether on others or for natural resources, as a cognitive eco-engineering function.

Some skeptics like Marco Viola (2017), however, argue that humans merely project first-person affective states onto animals. How could animals have moral emotions that motivate them, skeptics ask. One can never truly have first-person accounts of nonhuman emotions, Viola says. Laurent Dubreuil and Sue Savage-Rumbaugh (2019) who dialogue with chimpanzees and bonobos would disagree. While Viola might be factually true, he ignores the detail that, in line with Eliza Bliss-Moreau (2017), animals do not inhabit human niches, so their emotional ingredients and behaviors evolved differently. He also ignores the work of all those humans who signed or otherwise communicated with apes to grasp their inner feelings. Besides, by Viola's reasoning, does this mean, by implication, that one human can never completely comprehend the psychology of another human? We are using the wrong measuring cups (i.e., "human" and "nonhuman"). Rather, across humans and primates there are homologous emotions or evolutionary continuities. This means that structures but not necessarily functions are similar, whether a hand/foot or an expression of affection. Yet complex brains have shifting, merging, and contradictory motivations. Meantime in the United States, corporations are considered legal persons, and there has been talk circulating about when robots might be considered persons, yet those "bodies" do not exhibit or possess mammalian emotions or feelings. Programming computers or seating boards of directors on corporations does not spark them with personal or social motives to act.

The organic group dynamic, on the other hand, including all the networked brains, helped solve ecological problems like communal tasks via learning. This is not group selection but a group-level fitness equation. Natural selection happens on an individual level, but individuals constitute and could benefit the group. There is no group consisting entirely of cheaters and deceivers, since they would not be able to sustain themselves. There is a high degree of social intelligence, stemming from an individual for her own benefit, transferred to the group. So, contrary to skeptics like Viola, we can see the expression of ape feelings through their group interactions. An organism does not need a large brain to survive; see, for example, many dinosaurs. But one needs a large brain for socialization, which includes cooperation, collaboration, competition, and ecologically "conscious" food sourcing. Social skills take time and need to be learned, so a larger neocortex, involved in planning and decision making, is helpful. As with humans, this is true of great apes. Is the moral individualism of a great ape, say the young adult Harambe discussed in chapter 1, different from that of a human, say the young boy who entered the enclosure or the adult caretakers as a group? What motivated these different actors? Consider the moral individualism of all the great ape people like Washoe (with Fouts), Koko (with Patterson), Chantek (with Miles), and Kanzi (with Savage-Rumbaugh) who have certainly demonstrated what I'm referring to as an ape ethic, their motivation to connect socially with others and surely with forests had they been free in their home habitats.

While Robin Dunbar and Susanne Shultz (2017) admit that ecology, what to forage for, how and when, contributed to primate brain development, along with genetic development and parenting, they rightly lay emphasis on behavioral competencies like bonding and socialization. Of course my theme is that for apes there is a quasi-social attachment to the tropical bio-network, as well. Much psychic energy is involved in keeping track of who does what to whom, though I'd add, as I've been arguing, that the dynamic tropical forest itself plays a "social" role in creating the lives of ape persons. Specifically, consider here the neocortex, the most recently evolved brain part and the one involved with higher order brain functions and motivations. There is a correlation between brain/group size in primates but not in other animals. Though there are human/nonhuman primate brain similarities on a superficial level, on a deeper level there are cognitive disparities that seem to stem from a cellular reorganization of neural circuits in the human brain (Sousa et al., 2017).

Xiang Li et al. (2018) speculate, furthermore, that torque in the human brain, a slightly elongated twisting in the left hemisphere not seen in chimpanzee brains, might explain why humans have syntactic language and apes don't. Chimpanzees don't need grammar and syntax and have evolved very well without what one calls "language." Theirs is a language of movement and physical interaction with plants and trees. Differently organized brain networks are likely the result of large group/culture coevolution among human populations, one could posit. That ignores, though, the smaller groups of human ancestral hunter-gatherers and even *Homo erectus*. Just as there is no single gene responsible for any behavior, there are probably a number of factors to be accumulated, as adduced in my book, when painting the story of enlarged brains and moral individualism in humans and apes.

Although it is true that humans are genetically similar to chimpanzees and bonobos, from whom humans split via a common ancestor, there are key differences. Three percent, or about ninety million DNA nucleotides, according to Francisco Ayala and Camilo Cela-Conde (2017, p. 49), differ between humans and chimpanzees. Importantly, there have been more genetic changes in the human brain than in the chimpanzee's, especially in transcription genes that switch other genes on/off. However, continuing with the words of Ayala and Cela-Conde, "what we so far know advances but very little our understanding of what genetic changes make us distinctively human." The supposedly big evolutionary advances, like bipedalism and language, could be simple adaptations that helped early humans gain a slight edge for survival. Every organism's evolution can be reduced to one motive, how to survive adaptively, implying reproduction for genetic inheritance, too. While great apes do not have language per se like humans or routinely walk bipedally, they too have survived, mainly as social ecosystem engineers. In other words, there is no sterling distinction that gets pinned on the human chest. Each species evolved differently with colorful toolkits of adapted ecological and behavioral motivations, the whole point of speciation.

It seems likely that our African australopithecine ancestors of 3.5mya engaged in slow bipedalism even before the changing climate decreased forests and expanded open savannas. While great apes do not habitually walk bipedally, they can. In fact, chimpanzees, and more often bonobos, will walk bipedally if both hands are busy holding something or otherwise engaged, as in a dominance display. Orangutans can walk bipedally and will do so among humans. According to Ayala and Cela-Conde (2017), there might be a correlation between bipedalism and other human advances. For example, by the time of *Homo erectus* there was fast-paced bipedalism, and this changed pelvic morphology (smaller birth canal) and hence how offspring were born. Baby humans have soft cranial tissue and were born immaturely, which resulted in the need for prolonged care by the mother and a longer cultural development. Nonetheless, great apes also exhibit long periods, years in fact, of mother-infant care that includes alloparenting or care by another in the group, socialization, and communication.

As Ayala and Cela-Conde suggest, on a large canvas in comparison to great apes, there's nothing really distinct about being human, either in physical development or social motivations. One key difference I keep harping on, though, is how ape communities evolved to be relatively small and ecologically sustainable whereas most modern human communities have exploded exponentially with some obviously deleterious effects to their environs.

SOCIAL TRADITIONS AND CULTURAL ETHOS

There are many similarities across humans and great apes, such as locomotion, but each species has adapted differently to its environment. Orangutans, for instance, spend nearly all of their time in trees. Human evolution is not linear as characterized in a popular illustration of a human form rising from that of an ape. Rather, characteristics, traits, and morphology emerged in a mix-and-match, fits-and-starts process of slow, adaptive change. Take, for example, a very old hominin species of circa 4mya, *Ardipithecus*. In this species we find a hand that is close to that of a modern human, but not the foot, where a human foot might be closer to a living African ape's. *Ardipithecus* had long arms and short legs like current monkeys and apes, but not humans. Almost holistically, Darwin (1859, chapter 13) is less concerned about certain "trifling" traits in terms of descent and more focused on the prevalence of those characteristics in modified forms across species, relatedness.

How does adaptation translate to other behaviors? Early on Verne Grant (1963) was writing about the origin of adaptations in cultural evolution. Culturally, Hominoidea, the classification that includes humans and apes, has morphologies for different adaptive purposes. Chimpanzees have an extensive repertoire of tool use/culture, even using tools in sequence, for example, to obtain food resources like termites or honey. The fact that different groups of chimpanzees in different locations use tools differently points to cultural evolution. Personhood is social identity, and there are many ways of constituting such. For example, Hannah Mosley and Michael Haslam (2016) assert that idiosyncrasies associated with knowledgeable use of tools constitute personhood. Tool use is an extension of the body and mind and helps form an image of the self in the world. While there's a shared tradition of tools, personalities constitute a culture. Mosley and Haslam note that beyond any utilitarian function of tools, more important is how one and others perceive the tool user. My chapter 2 suggests tools as means of communication, and here the suggestion is that tools could serve a social function. Both perspectives are compatible. Some scholars (Peter Kappeler et al., 2010) insist on uniqueness for human and chimpanzee species, but that's a veiled attempt to claim human specialness. Indeed, one section of the introduction to their edited volume is entitled "Uniquely Human." What they claim as uniquely human—for example, cumulative culture, subsistence ecology, slow development and long lifespan, collaboration, planning, and communication—are traits evident in some forms across great ape species. These are quantitative and not qualitative differences in cultural traditions. Instead of unique species, we should be talking about moral individuals among all great apes, including humans. Barbara King (2004) speaks about the supreme usefulness of sound qualitative methods since they more intimately reach to the nuance of individuality and behavior.

In their analysis, Sue Taylor Parker and Michael McKinney (1999) demonstrate that the origins of intelligence don't begin with *Homo* but extend back into deep evolutionary time and are shared more recently with great apes. Nonetheless, with humans there has been a paramorphosis or overly extended development (life history) of cognition and intelligence. This is social, too. For example, concerning physical cognition, great apes develop such knowledge similarly to humans though at different rates. Apes have logical and mathematical abilities like those of a human child, though they develop more slowly. Contrary to what some thinkers suggest, namely Michael Tomasello and Josep Call, Parker and McKinney say that ape social cognition arrives across stages similar to those in human children. Apes engage in playful pretense socially at a very young age and thereafter in the wild, or as ex-captives preparing for forest reintroduction, as part of their cultural development. Likewise, Parker and McKinney say that great apes in the wild might be using symbolic communication for social purposes. There is chimpanzee tree banging, where the hits denote signals about travel direction or rest duration. Bonobos drag branches a certain way to direct travel and even leave a branch so later members can follow. There are reports of chimpanzees using leaf manipulation and display to initiate a grooming session or courtship.

Culture is not equivalent to "civilization." Much human culture can be ephemeral or even for its own sake. This is in sharp contrast to culture in animals, which is functional and can be passed down. Ape culture is grounded in a forest ecosystem and not in an artist's atelier, a musician's concert hall, a chef's kitchen, a fashion designer's studio, or a capitalist's factory.

Is there a huge gap between human culture and its equivalent among great apes? Andrew Whiten (2010) says no. Certain cultural features are shared between humans and apes, especially chimpanzees, and so there was likely similar culture in our common ancestor. For instance, says Whiten, in this shared ancestor there would have been, evident today across humans and apes, implements and social traditions as well as imitative and especially emulative learning behaviors. From these basics evolved more complex human and ape cultural practices. Culture consists of multiple traditions, seen in humans across the globe. Some other species simply pass on a distinct cultural practice, like subsistence foraging or song. Yet some species do show variation, for example, chimpanzees in disconnected regions with the same resources who crack nuts differently. Advanced and persistent cultural behaviors of the sort outlined here as part of a community point to personhood.

I don't disagree with Whiten in principle, but clearly there are vast differences in the outcomes of human/ape culture. Whereas human culture has helped its populations expand and consume goods generating incredibly dangerous waste, ape culture is eco-friendly and sustainable. For the sake of argument, though, let's continue with how both species create, possess, and transfer "culture."

In terms of human cultural evolution, a twelve-year study of chimpanzee extractive insect technology and percussive tool use for food processing across twenty areas was conducted by Thurston Hicks et al. (2019). Patterns of chimpanzee tool manufacture and use from sticks and clubs to stones imply diverse behavioral realms of traditions and toolkit assemblages across defined geographical regions. Tools include deliberately long and short probes for ants as well as dipping wands and sticks, so tools decidedly vary in length and thickness depending on purpose. There are also scoopers (for ants) and hammers (not only for fruit but for snails or tortoises). All of this comes with some regional variation but might be the result of expanded populations interconnecting, creating a combination of genetic matter and cultural material.

As William McGrew (1998) says, describing culture as uniquely human is "epistemologically suspect." Great apes have culture, and all primates share a common ancestor, ergo forms of proto-culture existed long ago and, as now, are likely social in nature. Culture is part learning, part phenomenon, and part adaptation, says McGrew. Chapter 2 of my book alluded to how Japanese macaques wash their food. While these are monkeys and evolutionarily older than great apes, let me relate that story, which dates back to Koshima Island, 1953. Then and there a wild macaque named Imo spontaneously washed a sweet potato routinely furnished by a human. Over time, the potato washing behavior spread and became a traditional norm. Then, in 1956, Imo placed wheat grains on the surface of water to remove sand. This practice, too, was subsequently copied. These behaviors are not static. The potato washing occurred, first, in fresh water and, second, in salt water with prolonged dipping. The monkeys across the group, therefore, were intentionally flavoring their food. Some of the monkeys dug holes near the sea so they could fill them with water and clean their wheat without fear of theft. There are other instances of non-subsistence culture, like the Japanese macaques who bathe in hot springs or collect and pile small stones or pebbles, says McGrew.

Nonetheless, some research by Sarah Marshall-Pescini and Andrew Whiten (2008) question whether chimpanzees are capable of cumulative culture. Their tests with young chimpanzees reveal that the subjects get locked into a stage of learning a technique for honey foraging beyond which they do not develop. The tests reveal that older chimpanzees quickly learn the honey probing technique that involves simultaneous ambidexterity and offers greater yield. With the younger chimpanzees, only one in eleven opted for the advanced technique. The authors do not discount ape traditions or basic levels of cumulative culture, though. Keep in mind, as mentioned, how the basic, early *Homo* toolkit remained steady for hundreds of thousands of years before improvement. Useful traditions stabilize in advance of significant innovation. Stability might also be a result of small groups, where variation would occur in larger populations.

While ape culture, which is more to bodily sustenance and forest maintenance, should not be measured against human culture, the fact is that there is "culture" across species. In fact, cultural behavior in apes, as with humans, can also be quite arbitrary, as in pop-up fads and transitory fashions. Take, for instance, the observed behavior of chimpanzees with grass placed in their ears. This began, say Edwin J. C. van Leeuwen et al. (2014), with Julie and her son who copied, and then others who had at least average association with Julie in a northwestern Zambian sanctuary. There is no adaptive value to this socially learned behavior, and it spread within a specific group of about ninety-four chimpanzees consisting of four different social groups. What is significant is how the grass in the ears is spontaneous, not based on reward, not functional (as a tool), not social (as grooming), and yet not caused by genetic or ecological factors. This is, then, an example of self-decoration in cultural evolution. How persistent will this practice be as a tradition?

Whiten (2010) is convinced that ape cultural practices, whether in tool use or courtship displays, are not just genetic or environmental but socially transmitted. He cites, for instance, the stone handling among the Japanese macaque monkeys. This practice seems to be an instance of cultural transmission since there is no function for the behavior. Primates like to handle stones for defensive purposes, so there might be an inherent psychological value to this nascent tradition. Can apes, therefore, ratchet up culture? Whiten cites the example of Goualougo chimpanzees who fish termites, first, with a stiff stick to dig a hole and then, with a pliable wand to collect the termites. This is controlled behavior that could not have developed in one step. Where a human child imitates and over-copies, a chimpanzee or especially an orangutan will engage in selective acquisition of knowledge and skills through close imitation and social emulation (Whiten, 2009). Imitation, socially driven as if to seek conformity, is not so easy and implies complex cognition in identification, understanding, and context (de Waal, 2001). There's an advantage, then, to such high costs.

There is ample evidence of orangutan culture, too, say Carel van Schaik et al. (2003). Culture can be a variant over ecologically similar geographies when, as in a custom, most individuals in the population demonstrate the behavior or when, as in a habit, the behavior occurs in at least a few socially important individuals. For orangutans, cultural variation is manifested in tool use, feeding, social signals like kiss squeaks, and other vocalizations similar to deviations seen across chimpanzee group sites. Biruté Galdikas and Gary Shapiro (1996) tell how the orangutans of Tanjung Puting engage in snag crashing, where a female or subadult male will deliberately send old vines or limbs smashing to the ground so as to mimic the motions of an adult male before his long call. There might be some eco-engineering or house cleaning in this behavior, as well. Moreover, say Galdikas and Shapiro, the free-ranging ex-captive orangutans at Camp Leakey would mingle and interact with the humans, and all types of accoutrement human, by choice, demonstrating bi-culturalism. Chimpanzees and orangutans, says Carel van Schaik (2003), competently exhibit cultural elements that include socially identified food preferences, signaling, and tool skills. These cultural evolutions are prevalent in humans, too. The only major difference is that humans routinely exhibit symbolic behavior. However, van Schaik and his team assert that orangutan material culture radiates "human like skill."

My question is this: Since intelligent and creative cultural traditions are evident in apes, can't we say that much of that constitutes their personhood? A related question about those persons concerns what purpose the culture and traditions serve. Do they enhance the ecological welfare of the species in its environment, or are they dangerously self-serving and environmentally unsafe?

SOCIAL LEARNING

If there is any section of my book where I seem to skirt very closely to human-ape similarities, it will be here, in my discussion of learning and culture. Readers need to keep in mind, however, that there is a prominent difference. Apes learn in forest environments and humans learn in buildings located in villages, towns, and cities. Apes are true ecologists, not modern humans.

Culture is social learning and can induce behavior beyond the explanation of ecological or genetic response, though genes and especially ecology factor in, as I've discussed throughout. Tests on captive chimpanzees provide evidence that experimentally fabricated customs can be passed along socially. Although important, ecological or genetic factors alone are insufficient to explain some behavioral differences between communities that are clearly culturally transmitted. Thibaud Gruber et al. (2009) conducted a field experiment on two genetically similar groups of chimpanzees in Uganda. There was the same physical task of extracting human-placed liquid honey from holes drilled in a lying, not standing, log. One group, known as stick users, did use sticks for extraction; the other group, known as leaf users, employed fingers and some leaf sponges. This means that when presented with a new challenge, wild chimpanzees fall back on their cultural knowledge. The implication is that behavior can vary outside of genetic similarities.

Like other primates, orangutans exhibit a high degree of neophobia (fear of anything new) in the wild, but not in zoos, say Carel P. van Schaik et al. (2016).

Tested out of captivity in Borneo and Sumatra by van Schaik, wild orangutans revealed little curiosity, unless socially prompted. The researchers repeated their tests on zoo orangutans who approached novel items (e.g., food in nests) quickly. There is a dependence on social learning and reluctance for exploration, even among adult orangutans. This is not to say orangutans are not innovators, but any new change, well maintained, is passed on socially. In contrast, orangutans in zoos are extremely exploratory, perhaps because of human contact and even boredom. Van Schaik says that orangutans in captivity have a far lighter cognitive load, knowing they are safe from predators with ready access to food. The main point here is that like other apes and humans, orangutans are social learners. Orangutan infants closely watch the physical and social behavior of their mothers in order to learn what to do and how to act. There is less cost and more advantage in relying on information from others or copying another's behavior and, perhaps, innovating from there. Of course some innovations are not socially induced and can come by accident, new movements, or necessity.

There is some artificiality in separating intelligence (chapter 2) and social behavior (this chapter), noting that in the ape lineage there was a coevolution of cognition and sociality since both support ecological networking and hence inclusive fitness. For example, bonobos, according to Jingzhi Tan et al. (2017) demonstrate xenophilia or prosocial responses toward members of other groups. Through experiments, Tan shows how bonobos will offer spontaneous assistance to an unfamiliar, non-group member in food acquisition absent any request for service. Xenophilia, says Tan, is "unselfish, proactive and automatic." How is this adaptive? In a social species, the advantages of making new connections outweigh costs, especially as aggressive behaviors diminished while strangers were accepted into the group. Testing tolerance, bonobos share all types of food resources often, even with strangers from neighboring communities. This openness reveals that hunting and meat sharing are not exclusively chimpanzee or human behaviors, say Barbara Fruth and Gottfried Hohmann (2018). Trust is not just human and is common because of shared evolution or convergence. Fundamental here is how establishing new contacts enables culture and cooperation to expand. Bonobos are not territorial, do not engage in infanticide, and demonstrate no lethal aggression against other groups. There is conflict, but severe aggression is unlikely. These behaviors are in contrast to chimpanzees, but both species have successfully evolved through social learning. While there is sexually selected infanticide in chimpanzees, gorillas, and humans, there are no known cases in orangutans, perhaps because of postnatal infertility, a fair amount of paternal certainty, and widespread ranging (Beaudrot et al., 2009).

While teaching is considered rare among chimpanzees, Stephanie Musgrave et al. (2016) demonstrate that tool transfer is a form of teaching.

To qualify as teaching, there is cost to the instructor, and the learner, who gains an advantage, must be present. This is the case in the Goualougo Triangle, Republic of Congo. Here, social learning involves the use of termite fishing probes, an adaptive behavior in the group across generations and transmitted, culturally, by teaching. Various tools are used to penetrate nests and then probe or fish. In some cases, a tool is used to puncture an underground nest. This combination of employing different tools for different tasks reveals complex cognitive strategies, and most teachers are mothers with infant offspring as learners. Tools can also be used socially, as in the case of an older female chimpanzee, Noel, who was observed cleaning the teeth of a deceased male, Thomas, whom she adopted as a youngster a few years earlier. Edwin J. C. van Leeuwen et al. (2017) say this behavior, from Zambia, was performed among the group who observed the action. Although most left the scene, Noel's adolescent daughter remained by her side to witness this extraordinary death response of acknowledging an individual.

Van Leeuwen et al. (2016) share further observations about the death of Thomas and the group members' responses. It seems, from careful review of videotaping of the scene, that there is a strong social component influencing a member's response to the death of another. When an infant dies, not many members of the community, other than the mother, pay heed. However, in this case, Thomas was socially active in the group and well-liked by at least one other male, Pan, a foraging partner. While there was quiet attendance around Thomas' body and careful inspection by many, Pan clearly was the most affected by the death, engaging in display, prolonged attendance, and more close inspection than any other member. Individual connections to close associates clearly mediate social emotions, partly a learned behavior. Amy Porter et al. (2019) report similar death concept behaviors in gorillas, including affiliative, investigative, and some agonistic responses. Those who had close social relationships with the deceased spent the most time near corpses.

Primatologists make clear that great apes have distinct personalities and recognize each other as individuals. Sociality is driven by individuation. One only needs to read field narratives by prominent primatologists like Jane Goodall (1971), Dian Fossey (1983), and Birutė Galdikas (1995b) or Goodall's academic work (1986) to see how these great apes evince particular character traits and temperaments. Evidence of this individualism abounds in how females of many species can be choosy about whom to mate with. Is that only biological instinct? Although subadult orangutan males actively seek females during the fruiting season, females will discriminate among the males and maintain spatial contact with an adult for protection, a social tactic (Fox, 2002). Learning depends on a variety of skill sets across individuals in a group, even among human persons.

Is there some rational component to personhood? A key question concerns the autonomy of an agent to act with some type of reasonable ability and how such autonomy is a freedom to be recognized. Being a person means having consciousness of a self and knowing that others can act differently. Going back to Wolfgang Köhler (1925), and reinforced since, we know that great apes deliberate, consider options, and then make decisions as outlined in many of the preceding sections of this book. They do not act only on primal instincts. Humans do not always act rationally but have instinctual responses—think about this the next time you experience fear, anger, jealousy, or sexual arousal. As group dwellers, these instincts, emotions, and intelligent choices are all moderated by social learning or how, for example, some feelings can develop into deep affection. Like humans, apes can be altruistic and compassionate in cases where there is little to no advantage gained, as shown by Tomasello (2014) and Warneken (2013). My argument, demonstrating ape/human differences, is that social learning for apes revolves around their ecosystem engineering as moral individuals evincing a land ethic.

Although Thomas Hobbes was essentially incorrect, for humans do not need a social contract under a strong leader to be civil, evidenced by long survival without such a contract, he nevertheless hints at how humans are a cooperative more than a competitive species. Anthropologist Sarah Blaffer Hrdy (2006) says that our cooperative breeding is what distinguishes us from our nearest relatives, the great apes. This too is somewhat anthropocentric since we tend to cooperate with other humans and compete with other species, for example, our systematic destruction of ape habitats (which are also home to some humans) and in consequence probable extinction of the apes themselves. Ample evidence of cooperation exists across the spectrum of human creativity, but particular attention should be paid to the arts, a form of communication using theory of mind. Chimpanzees and human infants alike can draw, but is it only the human child who comprehends that the drawing has symbolic meaning and significance? The definition and application of human personhood might hinge on our innate ability to manufacture and understand symbolic culture socially learned that feeds from and helps build human minds (Tague, 2018). But that anthropocentric designation should not deny moral individualism to great apes whose emotional understanding and learned behaviors enable their ethic for community and habitat sustainability through communication, tools, and culture.

As with the pioneering ethologist Konrad Lorenz (1965), the consensus today accepts that learning comes from adaptive modifications of behavior. This view does not exclude the importance of genetic inheritance and disposition. There's a correlation between the complexity of an adaptation and the small likelihood of its random change. Learning mechanisms evolved to exploit survival, Lorenz would say. Among apes, including human apes, learning is social. This means there's no dichotomy between innate or learned behavior. Only a blueprint is inherited, since behavior is not 100 percent innate, which explains why learning is vital. As recounted here, many apes have been absorbed into human culture, which implies that these human facilitators have been exposed to and to some extent have, consciously or not, adopted aspects of ape culture. Laurent Dubreuil and Sue Savage-Rumbaugh (2019) write about this ape-ape cultural exchange. Behavioral patterns from blueprint information are subject to development. Does this mean that humans could learn how to be better ecosystem engineers? A neural network enables the correct innate information, Lorenz might suggest, filtering stimuli and offering an appropriate response, as in maternal care or altruism. For apes, the inborn data shaped by social learning becomes their land ethic. The innate precedes and surely enables learning, what Frans de Waal has called our inner ape.

Can what is said above via Lorenz imply moral learning? An individual learns how to decode genetic information in social and ecological environments. Lorenz directly says that learning produces the adaptation of behaviors. Human "morality," in light of the evolutionary phylogeny (time) and clad (relatedness), depends on how genetic information is programmed and then processed through learning and culture. This process reflects the evolution of innate tendencies and modification of social behavior evident in great apes. If humans are persons, so are apes as morally individualistic inhabitants of a *Homo* line from a shared, common ancestor. A decisive difference and the refrain of my book, though, is how apes, cultivated in forest homes, are the ones who learn how to be effective ecological engineers and then pass that survival strategy on. Their evolved attitude toward sustaining the land is ethical.

OUTER AND INNER FOREST WORLDS

How Do Socially Learned Culture and Traditions Manifest for Apes?

Just as chapter 2 fortuitously ended with a note about gestalt, this chapter concludes with some thoughts about *Umwelt*, which already made an appearance. Biologist Jakob von Uexküll (2010) pioneered the notion of *Umwelt* in the early twentieth century, and the concept hinges on how matter interacts with organisms and how organisms perceive the world of matter differently. Readers should recall, too, Louise Barrett (2011) and the discussion of environmental psychology from chapter 2. The questions of perception, sensation, and comprehension have previously entered my argument with philosopher Thomas Nagel, to be addressed once again, below. The point to be made here, however, is that apes not only have a social system but one that can be termed as socioecology (van Schaik & van Hoff, 1996), especially among orangutans, where social organization via ecology and species biology are interconnected in their community. Most modern human societies are not, at present, socioecologist, but considering the responses to global warming, that could change. The last three "social" sections, then, prepared for this section in that the ape ethic is predominantly social ecology where individuals, partners, groups, or communities sustainably engineer ecosystems.

For purposes here *Umwelt* and to some extent gestalt both deal with how an individual experiences her physical and social environments. *Umwelt* is a sense of where one's self is at any given place and time. *Umwelt* is embodiment, including the subjective world of stimuli and gestalt. *Umwelt*, then, is an organism's life-world, and that can differ dramatically among animals. Canines live in a realm of smell. Bats live in a sphere of echolocation. Orangutans live in the visual horizon of a tree canopy environment. Many modern humans experience the world symbolically and mostly indoors. Consider the differences across human/"animal" species of actual contact with nature.

Umwelt is, according to philosopher Thomas Goudge (1967), an "environing world" determined by stimuli in an organism's sign-world and responses to its action-world. This means that organisms orient themselves in a self-regulating manner, a cycle of behavior where stimuli are selected for reactions based on their influence capacity. Contrary to Nagel (1979), Goudge suggests that through Uexküll's thinking a human can indeed, because of her own life-world experience, infer the *Umwelt* of another organism. Correlating to *Umwelt* is *Innenwelt*, or how an organism directs itself internally for environmental orientation and adjustment. In some ways, then, the natural world consists of infinitesimal gestalten interacting with *Umwelten*, or what James Lovelock (1979) refers to as Gaia. Humans could be, as they once were, part of this natural, communal interaction of self-regulating organisms but have mostly disconnected from it. Since there is no real but only a disguised division, a weighty ethical burden rests on preserving the biosphere.

Great apes are not aliens to the material components of their physical world, from the soil they sometimes eat for minerals, to the sky they touch at the tops of trees. Their bodily ecosphere is intimately part of their social and cultural worlds, not separate. Humans have managed to alienate from the natural world with their cityscapes, but a return is more than conceivable. At the same time, humans need to respect, therefore, the sovereignty of apes in the environment they have learned to carefully sustain. As I've demonstrated, forests for apes are inwardly psychological as much as outwardly earthly.

According to Uexküll, who did not exactly admire Darwin, an organism fits into its *Umwelt* because the organism understands and functionally responds to somatic and community stimuli and mental representations. In other words, there is semiosis or meaning making, which spreads among and across organisms, like cultural or social evolution. As with gestalt, in *Umwelt* animals perceive and act on needs to shape their future, indicating a sense of development in time. The holistic approach of *Umwelt*, where an ape mentally constructs her total physical and social reality, helps the argument of this book as I've shown in the long chapters on cognition and behavior.

Great apes do not live in an entirely random, mindless world but are active participants in the creation of their physical environments on both the individual and group levels. The shapes, contours, tastes, colors, sounds, textures, and odors of an orangutan's world are assembled by perceptual attributes and organized into a whole so that memory traces connect and associate. For example: desire/vision of fig; location of tree; climbing, grasping, and tasty nourishment. These symmetrical bundles of sensations enable an ape to achieve self-actualization as a moral individual in her private and public spheres as she relates with and communicates to her offspring, adults, and mainly the forest. An important difference here is how an ape in *Umwelt* is a more sustainable ecosystem engineer through gestalt than most modern humans. As I hope I have presented, in line with the thinking of phenomenological philosophers, tools are an extension of the body for coping with physical and social needs. The ape body is its tool in the forest. There are sensory and motor skills in a body schema that arrange movement according to tasks where space is not necessarily measurable but situational. This means, as per Hutchins (2008), that cognition cannot be reduced to a functional capacity or brain process. Ape cultural practices are part of a larger distributed cognition rooted in an active ecosystem.

Rowlands (2019) sees causal reasoning in animals (e.g., corvids). Logical reasoning, a bit higher on the scale, is basically along these lines: either A or B; not A; therefore B. That is, one has a belief based on logic and not simply mechanics. We can see both causal and logical reasoning in apes related to culturally social tool use. Great apes "know" that if they learn how to resource food using tools or by manually manipulating fruits and leaves their action necessitates a change in the food source, that is, disappearance with probable replenishment. They don't need to know how this "works," just as most people don't know how a car engine works. Ecosystem engineering, then, seems to be a blend of causal and logical reasoning based on inherent, learned, and socially influenced behaviors. This assessment is crucial since, to borrow from Rowlands, it grants psychologically moral or social agency to apes.

That is, in *Umwelt*, an ape maps a biological self in an ecologically responsible manner, consciously or not, configuring patterns of behaviors that interchangeably sustain the self in a biome.

To conclude this chapter, recall its main divisions and connections, in relation to the long tail of my argument, as I proceed to the Conclusion. For apes as much as humans there is biology in moral systems, with the difference that primates have an evolutionary ethics geared toward social conformity and wholeness in a bio-networked jungle ecosystem. In their sociality with conspecifics and, primarily, wooded lands apes exhibit attributes that can only be described as personhood, with traditions and a cultural ethos learned from intimate communal living and sustainable engineering in tropical forests. Theirs is a life of psychologically and physically connected outer and inner forest worlds.

Conclusion

Long Call for Ape Forest Sovereignty

There's a question of personhood for great apes because of their close similarity to humans. Yet a note of caution is in order. I explore similarities of cognition and behavior only to highlight differences. To wit, apes are better at maintaining vital forests and servicing climate health than are humans. In crafting this conclusion, and echoing some of what has preceded, I invoke the moral failures of humanity in areas such as animal welfare, sustainable land use, as well as cruel and destructive agricultural practices. If my tone here seems strident, so be it, for what I have to say comes from a place of honesty and authenticity.

My Introduction provided an overview of the vital ideas and concepts of this book. Now, I offer an overarching reflection on the various key claims and crucial material presented in the argument respecting the moral and ecological personhood of great apes that would grant them sovereign dominion over their tropical forests. Special attention, I ask, should be paid to the call for action evident in these final words.

Following some of the embodied cognition and ecological psychology from the above chapters, gastroenterologist Emeran Mayer (2011) says *gut feelings* arise through a relationship connecting the brain and enteric nervous system. That visceral sensation is enhanced in apes since they live in a thriving ecosystem putting their bodies in tune with nature unlike humans. I hope I've made clear that apes experience an emotional intelligence enabling them to maintain their habitats in ways forgotten or deliberately lost by most modern human societies. I trust that readers see why this land ethic of ecological engineering is fundamental in nourishing tropical forest biomes which, in turn, affect the fitness of the earth's biosphere. So while, at times, I cataloged ape-human similarities, the underlying message is on the morally individualistic ape ethic of sustaining tropical lands as habitable domains for themselves and other species with the consequence of global climate health.

THE LANGUAGE OF ANTHROPOMORPHISM

As judges Sharma and Sing (2018) remark, the opposite of person is not animal. In their decision for animal rights, these judges, quoting from the ancient Isha-Upanishads, remind us that humans are part of nature and don't have their own special rights over other species. According to the United Nations (1982) World Charter for Nature, uniqueness and worth appear in all aspects of the natural world and demand from humans a respect for the sanctity of life. Many countries now include animal rights protections or welfare in their constitutions or by legislation. Is this hopelessly romantic? No, and there's science to back up the long argument of this book for great ape personhood, as I hope I've demonstrated.

Philosopher Martha Nussbaum echoes primatologist Frans de Waal by saying that "human beings very often fail at moral compassion because of . . . the human tendency to deny its own animal nature." Similarly, philosopher Tom Regan (2004) suggests one need not question anthropomorphism. Really, there are not human-only traits. Animals have consciousness, experience a range of feelings (e.g., pain) and emotions (e.g., affection) like humans, and behave in ways (e.g., communication) similar to humans, especially primates and particularly great apes. Regan likewise talks of potential. For instance, apes don't speak like humans since they don't have the physiological apparatus, but they nevertheless communicate emotionally and symbolically with humans, as shown by Laurent Dubreuil and Sue Savage-Rumbaugh (2019) who have conversed with apes about simian motives, needs, and desires. Elsewhere, Regan (1993) says the human "constellation of values" is revealed in how some societies let people viciously treat animals. What are morally relevant differences between great apes and humans? Of course I agree with him that there are minor variances, but let's move this to the next level. Legislation to grant great apes sovereignty over their tropical forests, by virtue of the care and concern they express for those lands, is the direction, now, humans should follow.

APE/HUMAN MORAL PSYCHOLOGY

As Regan (2004) points out, the nineteenth-century English, utilitarian philosopher Jeremy Bentham argues that because animals suffer, in spite of their welfare and preferential interests, they should not be deliberately harmed any further by humans. Like humans, animals avoid not just pain but especially suffering. The more intelligent a creature, the more it suffers. Consider then the mental suffering of imprisoned great apes in zoos, road shows, and research facilities, to say nothing of other intelligent and social "animals" from rats to elephants. When people cause animals to suffer contrary to their own welfare, the moral weight is on human shoulders. In this respect, consider how primates across the globe are suffering because modern industrial farming and capitalistic agriculture are destroying their forests. The apes are shot and killed as pests. They are stolen to be trafficked as pets. They are caged and sold to zoos. They are burned alive.

In one Frankenstein experiment, pig hearts were transplanted into baboons (Baker, 2018). If humans are the most civilized, moral species expressing rationality, why would they inflict such terror on animals who likewise think and feel? Chinese scientists (Shi et al., 2019) have actually put human genes into the brains of monkeys which made them more "human-like" in terms of memory and development. One of the team's researchers has expressed desire to implant a so-called human-specific intelligence gene into monkeys (Leitch, 2019). Ape persons evolved as such by virtue of their moral intelligence and emotional behaviors, as I trust I've effectively exhibited in this book. Humans don't need to create a human-like ape psychology since great apes already have their own socially and ecologically sustainable ethos. As Barbara King (2017, p. 193) rightly says, "No animal need be sentient *like us* to be sentient, just as no animal need be smart like us to be smart."

Moreover, the use of the adjective "nonhuman" trivializes the feelings, emotions, and intelligence of many species, especially primates, who share with humans a history of morphology, sapience, and sentience. When the dying chimpanzee Mama was reunited with her long-ago researcher friend Jan van Hooff, as described by Frans de Waal (2019), one could see in all of Mama's movements and emotions quintessential human behavior: the greeting, embrace, smile, stroking of hair, and arm around Hooff's neck to embrace and pull him closer.

APE/HUMAN MORAL SENTIMENTS

Dale Peterson (2011) suggests that mammalian moral psychology from a deep evolutionary past is emotional and contingent upon cognition and intelligence. Attachments among individuals in a group and even between groups and different species abound in nature. As the historical record curiously shows, *Homo sapiens* lived simultaneously with Neanderthals. Why did modern humans survive and they did not? Some scholars believe people contributed to Neanderthal demise by taking over their domains, much as human societies are now doing to primate habitats. No doubt there were many causes for their extinction, but is it inconsequential that Neanderthals died off after the advent of *Homo sapiens* in Europe? Even Darwin knew that in the struggle for existence like species compete for the same resources in the same biotic community. At any rate, with their culture, Neanderthals were persons.

The pace at which humans are eviscerating the natural world is staggering, so let it stop at the edge of tropical forests. Expanding forests in Africa, South America, and Indonesia, homes to many primates and apes, by granting them land rights and forest person status is a step in the right direction. Science and technology alone are not necessarily the solutions for solving environmental problems, anyway. Sociologist Eileen Crist (2019) sees mass extinction due to the spread of destructive technology to produce food, energy, and commodities. An uptick in moral attitudes and social habits could do as much or more good.

For example, imagine the state of affairs if half of the world's population dramatically cut down on animal products, especially dairy. Some people like Simon Fairlie might argue for small scale meat farming. Scientific research shows that even small farms emit unacceptable levels of methane (Wolf et al., 2017), and grass-fed beef requires larger cattle populations, hence greater pollution (Hayek & Garrett, 2018). My point, of course, goes back to how great apes are good ecosystem engineers who tend sustainably to tropical forests without agriculture or pastoralism. Correspondingly, conservation biologist Brian Czech (2016) points out how some economists grossly neglect the science of unchecked population explosion, production, and consumption ecology. There are environmental limits to excessive economic expansion that devalue any steady state growth. Perpetual growth economy is outstripping matter and energy, Czech would say.

APES/HUMANS: NOT AS DIFFERENT AS IMAGINED?

In 1983 Roger Lewin asked if the orangutan is a living fossil. H. Lyn White Miles (1994) goes as far as saying how only a few genetic changes would theoretically separate all great apes, including humans, from an orangutan-like ancestor. Among two possibilities for hominid evolution, according to Lewin, are the specialized facial features of *Sivapithecus* and *Ramapithecus*, from the middle Miocene throughout Eurasia, shared with the orangutan. There is no ancestral link to other apes and humans. *Sivapithecus* is a forerunner of the orangutan and lived in areas now known as northern India and Pakistan about 8–12mya but became extinct at the end of the Miocene when forests died off to a colder, drier climate. Monkeys adapted. In another scenario where *Sivapithecus* and orangutans share a primitive face, there is linkage to later species, including humans. But as Boyd and Silk (2009) succinctly put it, the orangutan, who shares features derived from *Sivapithecus*, is the only clear antecedent of modern apes currently known. Controversially, Schwartz (1987) puts more focus on the fossil record and not on molecular similarities and so finds a human ancestor closer to the orangutan than to African apes. Other theories, especially those utilizing DNA correlations, point to chimpanzees and bonobos as our closest living relatives.

Based on dissections of a number of male, female, and juvenile bonobos, Rui Diogo et al. (2017) assert that the bonobo is the best model for any common ancestor to chimpanzees and humans. Male chimpanzees are aggressive, competitive, and dominance oriented. Male bonobos are not, and defer to females. Male chimpanzees will cooperate in order to attack other groups. Such alliances and aggression are unheard of among male bonobos. In fact, bonobos more than chimpanzees enjoy pleasure and amusement as adults, whether from sexual activity or simple play. The point of this catalog, according to Kay Prüfer et al. (2012) is to show how humans are more similar to both bonobos and chimpanzees than the great apes are to each other. This, along with genetic information, explains how human behavior is a variegated pattern from immediate predecessors, like the chimpanzee and bonobo, and more ancient antecedents, like the gorilla and orangutan. Indeed, Derek Wildman et al. (2003) have gone as far as saying that since DNA sequencing presents an objective, data-driven snapshot of different hominid genomes without the possibility of anthropomorphizing results, the only logical conclusion is that humans are "remodeled chimpanzee-like apes." The implication here is that the genus *Homo* should include, primarily, the chimpanzee and the bonobo, and secondarily, the gorilla and orangutan.

Other animals are clearly self-aware, most conspicuously by reacting to painted dots on their foreheads or bodies revealed in a mirror. This higher consciousness, first tested by Gordon Gallup (1970), is evident in chimpanzees, bonobos, orangutans, gorillas, and dolphins. As noted, but worth emphasizing, animals have individual personalities, and in the word "personality" one acknowledges a person. Personality is at work on an interdependent level. Humans alone do not have a division of labor, since male chimpanzees patrol and fight while females nurture young. These young, like human babies, require extended care and cultural training, including recreational games. In fact, many species sport and play as a format for youngsters to test and simulate their abilities in the face of opponents for competition and cooperation.

If there's any human/ape difference, it is centered in how intelligence and social behaviors are used by apes to sustain resources and habitats in small, controlled populations.

A CULTURE OF CHANGE?

Moral attitudes change over time because of advances in scientific findings and, especially, through cultural evolution. Peter Wenz (1993) in the 1980s rightfully argued for human moral obligations toward the ecosystem. I've tried to demonstrate that apes, as intimate, biosynergistic mechanics in their habitats, in turn positively affecting surrounding bio-networks, exhibit moral obligations socially and environmentally. Going backward, some authors like William Baxter (1993) actually argued for optimal pollution in the 1970s whereas now people talk about zero waste. Great apes have consistently practiced a life of producing biodegradable waste. While Peter Singer (1975) was ahead of his time in calling for equal consideration of human/animal interests, there have been dissenters. Carl Cohen (1993) denies that animals have rights and defends animal experimentation by putting humans atop an imaginary hierarchy. While Christina Hoff (1993) writing in the late 1970s admits that animals suffer, she contends that they can be used in painful experiments to serve humankind. I do not hold positions like Baxter's, Cohen's, or Hoff's and wonder how planet earth, much less "humanity," is sustainable with such confrontational attitudes toward nature and its inhabitants. To what degree have some of these attitudes changed?

Part of the moral and environmental problems we now face, as outlined by Sindhu Radhakrishna and Dale Jamieson (2018), deal with negatively aggressive human attitudes. These authors say that in Western societies, which have dominated the field of primatology, the emphasis has been on anthropology, or how primates fit into the human picture of history. Eastern societies, like India, Vietnam, and Japan (as well as those in Brazil) which have intimate contact with primates don't see much of a distance between the human and animal worlds. Could that work on a practical level where apes and humans coexist, as Donaldson and Kymlicka (2011) and Meijer (2019) suggest? Some authors say that if poaching and hunting of orangutans, for example, were to cease, these forest persons could be adapted to some multiuse human landscapes, as evidence shows (Spehar et al., 2018). Gisela Kaplan and Lesley Rogers (2000) asked a similar question, coming after the severe Indonesian droughts and fires of 1997 and 1998. Can orangutans coexist with developmental plans, that is, human architectural communities on the fringes of rainforests? The skeptic in me says probably not, since some influential humans, using power or money, will always want more land, lumber, or other natural resources and likely continue encroaching into the forest.

The problem here is that regrowth forests, as Kaplan and Rogers (2000) say, have poor soil because of weak root systems, not enough plant species, and certainly trees not old enough to reproduce properly or tall enough for orangutan living. This assessment echoes firsthand observation by Barbara

Harrisson (1962) who says that after human degradation, for example, tree cutting for a rice plantation, the forest balance is so upset that new growth is unable to restore harmony. Some species of these trees require 100 years to mature fully, but they are cut and replanted in cycles of about twenty-five years. That creates a black hole between what is cleared and the regrowth. This is not, contrary to what has been advertised, sustainable for orangutan life. So-called selective logging is no better since in order to get to the selected tree many others are damaged or destroyed by heavy equipment or flattening to create roads. Dead trees are left in abundance to rot, which inhibits valuable ground soils from achieving maximum density and even affects air dew points or high levels of moisture, critical for a rainforest. Likewise, reducing the size of rainforests to restricted preserves puts immense pressure on the area by a number of competing species and so negates orangutan eco-engineering since they naturally forage across large, overlapping ranges. Any benefits, then, in controlled forestry or preserve allocation fall to human consumers, not to orangutans. How do we ameliorate such waste and persecution?

There needs to be a massive attitude shift away from the human-centric. How else are great apes to achieve personhood? As Richard Byrne (1995) shows, human intelligence can be examined through a comparative analysis of the evolutionary psychology of great apes. If human intelligence is a key criterion for personhood, and since great apes exhibit the same types of intelligence in different degrees, then apes are persons. There's optimism, however, as more young people are aware of and sensitive to animal rights and conservation. They might ask questions like these. Do we need to eat meat? How much more palm oil do we need? Their voices are votes for a greener future that ensures the rights of all living organisms. This is especially true in the arts and humanities, where those voices are speaking out in creative ways and offering images and ideas to help redirect human culture as a positive force for nature.

Sociologist Lori Hunter (2016) says the overall human environmental footprint is huge because of enormous population dimensions, especially in some cultures where children are socialized to over consume products (USA) or to eat disproportionate amounts of meat (China). In other words, we don't need to invest in technology to solve some global conservation and animal rights issues. People need only to adjust their moral position to one that is animal kind and ecologically friendly.

This book accentuates how apes and humans share many characteristics, traits, emotions, and intelligence. However, I hope I've exposed a critical difference regarding ecological sensibility in how apes like orangutans are far better ecosystem engineers than most humans. At least taking up that attitude would be a big step for most people who currently view apes as "animals" and zoo entertainment and not as multiple hominid species who care for the crucial climate health of forests. There are many voices and advocates for conservation and animal rights, but let's turn to Marc Bekoff (2013), who summarizes anthropologist Barbara King. Very early humans lived and evolved in balance with many different animals in the natural elements, and this connection with all of the chemicals, ingredients, plants, and products of nature is ingrained into our ancient psyche. Animals for prehistoric people were not just something to eat but organisms to share space with and live next to. All creatures in prehistory lived in green spaces, but with large-scale game hunting followed by agriculture and pastoralism, and then the rise of cities along with technology and industry, most humans lost or broke any link with the natural world.

THE GARDEN AND ITS GARDENER

At the end of *Candide*, Voltaire enjoins people as follows: *Il faut cultivar notre jardin*. We must tend to our garden against certain ills of the world. The emphasis here, as noted by literature professor Robert Pogue Harrison (2008) is on *notre jardin*—our garden, a collective and not a private enterprise. The human conception of perfection, says Harrison, includes the vision of a garden. Indeed, when one considers the opposite think of T. S. Eliot's modern nightmare *The Waste Land*. In contrast to any poet's existential angst, primatologist Biruté Galdikas (1995b) entitled her study of orangutans *Reflections from Eden*. Contrary to Harrison, however, gardens are not merely the domain of what he calls the human condition. Contentment is not in abstract musings but in the satisfaction of one's body in nature. Todd Koel et al. (2019) studied forty years of data on a wildlife area revealing how one invasive species, or an "apex predator" perhaps introduced by modern humans, had negatively cascading effects on other species. In an otherwise pristine area there were devastating results for species from plankton to fish, birds, and mammals. My question is this: In terms of widespread tropical forest decline, who is the apex predator? In a like manner, which species has the self-discipline not to ruin the land it inhabits?

In line with Adam's curse, humans must labor and often confuse necessary work with unnecessary consumption. Before the Fall, mythical Adam and Eve were self-realized vegetarians and lived to care for their basic needs through the garden, much as do apes. Not without toil and pain, apes, however, seem more mindful of their forest gardening than humans, as the evidence of *An Ape Ethic* supports. There is clearly a different ethics of caring between the handiwork of a great ape and a human gardener. Apes provide trees room to grow, where plant and animal life generate more life, as opposed to human forest clearing and the application of pesticides and herbicides. Have humans truly adapted to plant species or have they, rather, "engineered" them artificially through industrial means for production or aesthetic purposes? In the case of pollination, by analogy, Darwin (1859, chapter 4) says that a plant and an animal can modify and adapt to each other. That has been my key point about nonhuman primate eco-engineering, which is closely linked to the ape ethic of moral individualism and their expression of duties and obligations to others in their forest habitats. Not only do apes live in an ecological bio-network, they have created and sustain it.

To read authors of the Middle Ages and the Renaissance, one finds small, manicured, walled gardens attached to a house. The garden is usually locked from the outside world. The etymology of paradise suggests surrounding walls. See, for example, the Merode Altarpiece (in the Cloisters collection of the N.Y. Metropolitan Museum of Art) where the garden is part of religious iconography. Why do humans spiritualize gardens and, therefore, separate them from the simple facts of and connections to nature, bugs, worms, weeds, and all. For humans, a garden is a place to impose artful order. Gardens, then, become more like cultural figurations of status and less as places for natural growth. While a human gardener will often think about his plants and intellectualize his activity, not so for the orangutan gardener. This is why it is difficult to measure the ape land ethic of moral individualism because the criteria vastly differ. But the evidence is in plain sight. An orangutan is in constant contact with a garden which does not exist independently of her or nature. In line with Epicurus, salient ingredients of patience and acceptance are critical for a healthful physical and mental disposition. Which gardener exhibits these traits? Is it the human who justifies his savage destruction with abstract notions like biophilia or the orangutan who in spite of her appetite exhibits the paramount moral virtue of temperance in biosynergy?

By the sixteenth century in Italy and the seventeenth century in France, aristocrats literally moved large tracts of earth to create elaborate gardens symbolizing their wealth and power. These private enclosures were on a huge scale representing ambition and politics, the domination of wildlife through human imposed order. The grandest of all these gardens, Versailles, was set deliberately as a museum of plants, a static diorama of fashionable designs where space, textures, colors, and objects were manipulated. With gardens, no matter how large, there is an unmistakable divide between inside/outside with the display of control. For many humans, the garden is a performance tableau. In a bizarre cognitive twist, it's as if humans have confiscated and assaulted tropical forests as massive, toy gardens. By now you see that for apes, on the other hand, their horticulture and pruning are more eco-friendly and sustainable than what is "engineered" by humans.

Harrison (2008) quotes Saint-Simon regarding the gardens of Versailles: *ce plaisir superbe de forcer nature*—the prideful domination of nature. The human garden is enchanted, artfully shaped, a science fiction of sorts, and indeed a science experiment. The orangutan garden is reality, perpetual, and shared. Whereas humans engage in existential nihilism, orangutans are fulfilled in the caretaking, not the *taking*, of the garden's fruits since one's *domus* or home is the garden and not an economic venture or a separate place. As Darwin (1859, chapter 4) says, natural selection acts for the preservation of advantageous variations. With anthropogenic stressors on forest habitats, this means human beings are reversing the course of nature. Darwin is right to assert that great diversity supports all life. To hack away at large chunks of biological variety which supports the whole over expanses of time removes cornerstones of what he terms the "polity of nature." Diversity and competition without walls encourage interdependency and growth in lockstep, so with human actions there will be a toxic chain reaction, a reversal of diversification in the garden.

For apes, the forest environment builds into and reinforces their ethos, whether in a savanna, rainforest, or mountain woods. The tropical forest habitat for an ape is like its own flesh. I'm not sure we can say most humans in modern times have quite the same indivisible union with the natural world. As Rowlands (2019) argues, to help my claim of ape ecosystem engineering, not all awareness needs to be intentional, and I believe this is discovered in how primates and other animals negotiate their physical environs. Because of their high level of cognition, apes might expect they are inclusively involved in their environment to make changes regarding fitness for self, offspring, kin, or others. Physical and mental states are not understood and do not occur in a vacuum. Bodily awareness implies action, and with a highly cognitive species, such action is imprinted upon and routed back into the species to cement close physical contact with nature. This is how I see apes as eco-engineers, especially considering the wide ranging salubrious effects of their thoughtfully deliberate foraging.

Human gardens are functional agriculture, says E. O. Wilson (2017). Gardens bridge the Paleolithic and Neolithic eras. Early people would have understood the physical and mental health benefits of a garden, says Wilson. Human gardens, however, are more to architecture than to agriculture, since they are often filled with non-native species and thus can act as a detriment to the local insects and insectivorous birds. Human gardens are also pits of water waste, insecticide, and herbicide that infect ground water tables. These deleterious actions or effects are not evident in great ape ecosystem engineering. Land conversion equals fragmentation and is forcing apes and other species to adapt to dwindling forests that abut agricultural areas.

As Rosa Garringa et al. (2019) note, this fragmentation adjustment is happening now, and in southwest Sierra Leone chimpanzees are competing with human farmers for resources. The chimpanzees who have lost forest homes to slash or burn agriculture are managing to survive in fallow fields, swamps, and mangroves. Where are their gardens? Who is the moral individual in this scenario? At a minimum, these apes need corridors that connect large, forested areas.

THE LONG CALL

This final section has been written in response to some of the challenging ideas found in *Terrorists or Freedom Fighters? Reflections on the Liberation of Animals* (Best, 2004). One of that book's foundational thoughts derives from philosopher Jiddu Krishnamurti. There are no human relations without a relationship with the natural world. The fate of one species is linked to another and then others in an ever widening gyre, as ancient thinkers saw, and as some modern writers have warned. That theme has been echoed throughout my book. People are now witnessing the adverse cascade of damage human induced climate change is spreading in suffocating rapaciousness across the planet. What's worth more, property or life? Corporations own property and have legal rights, but what about ocean and forest animals? At what cost are some humans willing to go, to irrevocably destroy stretches of plant and animal life on earth for a vacation home or a secure investment? Of course my point has been that much detrimental human activity is encroachment on sovereign lands held onto and tended by various animal species.

In a rainforest biome each species of plant and animal has evolved to service the whole breathing fabric and has a greater claim to life than the corporate oppressors who destructively invade. A false dependency has been established. Most humans do not need the wood, minerals, water, and especially not the trafficked "pets" of the rainforest. In the end, one standing on the perimeter needs to ask, *What makes sense? Who is the moral individual?* Rather than being mindful of the critical functionality the rainforest provides for the world's biosphere, too many people have become mindless. The natural world is objectified, fragmented, and abused, tactics historically used by powerful humans to enslave or systematically eradicate others. Now, insects, the world's pollinators, are vanishing. An unnatural distance has erupted between many humans and other natural forms. The few hundred remaining Tapanuli orangutans on Sumatra face almost certain extinction in the wild because of plundering in the name of social or economic progress.

I should note, however, that many of the researchers I cite in my book, from Jane Goodall to Biruté Galdikas, are working hard on different fronts.

Goodall is a worldwide spokesperson for animals and conservation through her organization, and Galdikas continues her field research in Borneo. People like Gary L. Shapiro, to whom this book is dedicated, and others like Leif Cocks and Willie Smits, to name only a few, work tirelessly on the ground to foster sustainable farming among local people, to rescue, rehabilitate, and reintroduce apes into the forest, and lobby to preserve, protect, and enlarge rainforest areas. There are also organizations like Gorilla Doctors and brave souls like the Virunga Park Rangers who literally risk their lives against poachers and other human forest invaders to protect gorillas. So there are many hard-working, morally minded, decent humans advocating for primates, apes, and other species. Yet, more people need to join these important causes by contributing as they can.

What is not often seen, since it is rarely reported, is the violence heaped upon protected areas of nature and animals for the sake of corporations, whether in agriculture or gas drilling, whose sponsor dollars line the pockets of lobbyists, legislators, and media executives. In a recent, astounding development, the U.S. government periodically considers using cyanide bombs to kill wolves, mountain lions, bears, foxes, birds, and such creatures that farmers and cattle ranchers see as pests. The call, then, is for a revision of values that shifts attention away from the perimeter of our attention and places much thoughtful consideration toward caring for the natural world, as evidenced in our great ape cousins, the moral individuals whose land ethic consists of efficient ecosystem engineering.

I pray I've convincingly demonstrated, then, that in some respects the differences between ape and human are more striking than any similarities. I spent much time cataloging human/ape similarities only to tease out contrasts. While all these species are alike in terms of cognition, intelligence, and social behaviors, the great apes are the good eco-engineers, not humans, since they have not systematically degraded lands as humans have. People might be able, therefore, to learn something about caring for the earth from our primate cousins.

Bibliography

Aaltola, Elisa. 2015. "The Rise of Sentimentalism and Animal Philosophy." *Animal Ethics and Philosophy: Questioning the Orthodoxy*. Elisa Aaltola and John Hadley, eds. London: Rowman and Littlefield, 201–218.

Alexander, Richard D. 1987. *The Biology of Moral Systems*. New York: Aldine de Gruyter.

Allchin, Douglas. 2009. "The Evolution of Morality." *Evolution: Education and Outreach* 2, 590–601. Online.

Allen, Colin. 2001. "Cognitive Relatives and Moral Relations." *Great Apes and Humans: The Ethics of Coexistence*. Benjamin B. Beck, et al., eds. Washington, DC: Smithsonian Institute Press, 261–273.

Altieri, Andrew H. 2015. "Ecosystem Engineers." *Oxford Bibliographies Online*. doi: 10.1093/OBO/9780199830060-0134.

Alvaro, Carlo. 2017. "Ethical Veganism, Virtue, and Greatness of the Soul." *Journal of Agricultural and Environmental Ethics* 30.6, 765–781.

Alvaro, Carlo. 2019. *Ethical Veganism, Virtue Ethics, and the Great Soul*. Lanham, MD: Lexington Books.

Andrews, Kristin. 2015. *The Animal Mind: An Introduction to the Philosophy of Animal Cognition*. London: Routledge.

Andrews, Kristin. 2016. "Animal Cognition." *Stanford Encyclopedia of Philosophy*. Edward N. Zalta, ed. Stanford: Metaphysics Research Lab. http://plato.stanford.edu/. Online.

Andrews, Kristin. 2017. "Chimpanzee Mind Reading: Don't Stop Believing." *Philosophy Compass* 12, e12394.

Andrews, Kristin. 2018. "Apes Track False Beliefs but Might Not Understand Them." *Learning and Behavior* 46.1, 3–4. doi: 10.3758/s13420-017-0288-8.

Andrews, Kristin, et al. 2019. *Chimpanzee Rights: The Philosophers' Brief*. London: Routledge.

Anscombe, G. E. M. 1958. "Modern Moral Philosophy." *Philosophy* 33.124, 1–16.

Aristotle. 1952. *Nicomachean Ethics*. W.D. Ross, trans. Chicago: Britannica.

Atkinson, Elizabeth Grace, et al. 2018. "No Evidence for Recent Selection at *FOXP2* Among Diverse Human Populations." *Cell* 174, 1–12. https://doi.org/10.1016/jcell.2018.06.048.

Ayala, Francisco J. and Camilo J. Cela-Conde. 2017. *Processes in Human Evolution.* Oxford: Oxford University Press.

Baker, Noah. 2018. "Listen: Pig-heart Transplants in Baboons Work for Longer Than Ever Before." *Nature*, 5 December. Online.

Bard, Kim A. 1990. "'Social Tool Use' by Free-Ranging Orangutans: A Piagetian and Developmental Perspective on the Manipulation of an Animate Object." *"Language" and Intelligence in Monkeys and Apes.* Sue Taylor Parker and Kathleen Rita Gibson, eds. Cambridge: Cambridge University Press, 356–378.

Bard, Kim A. 1992. "Intentional Behavior and Intentional Communication in Young Free-ranging Orangutans." *Child Development* 63, 1186–1197.

Bargh, John A. 1994. "The Four Horsemen of Automaticity: Awareness, Intention, Efficiency, and Control in Social Cognition." *Handbook of Social Cognition.* R.J. Wyer and T.J. Srull, eds. Hillsdale, NJ: Erlbaum, 1–40.

Barrett, Louise. 2011. *Beyond the Brain: How Body and Environment Shape Animal and Human Minds.* Princeton: Princeton University Press.

Barry, Joshua M., et al. 2018. "Pumas as Ecosystem Engineers: Ungulate Carcasses Support Beetle Assemblages in the Greater Yellowstone Ecosystem." *Oecologia* 188. https://doi.org/10.1007/s00442-018-4315-z.

Bastian, Meredith L., et al. 2012. "Innovative Behaviors in Wild Bornean Orangutans Revealed by Targeted Population Comparison." *Behaviour* 149, 275–297.

Baxter, William F. 1993. "People or Penguins." *Moral Controversies.* Stephen Jay Gould, ed. Belmont, CA: Wadsworth, 550–554.

Beauchamp, Tom L. and Victoria Wobber. 2014. "Autonomy in Chimpanzees." *Theoretical Medicine and Bioethics* 35.2, 117–132.

Beaudrot, Lydia H., et al. 2009. "Why Male Orangutans Do Not Kill Infants." *Behavioral Ecology and Sociobiology* 63, 1549–1562. doi: 10.1007/s00265-009-0827-1.

Begun, David R. 1999. "Hominid Family Values: Morphological and Molecular Data on Relations Among the Great Apes and Humans." *The Mentalities of Gorillas and Orangutans: Comparative Perspectives.* Sue Taylor Parker, Robert W. Mitchell, and H. Lyn Miles, eds. Cambridge: Cambridge University Press, 3–42.

Bekoff, Marc. 2000. "Animal Emotions: Exploring Passionate Natures." *BioScience* 50.10, 861–870.

Bekoff, Marc. 2015. "New Zealand Declares Animals to be Sentient, Bans Testing." *Psychology Today.* https://www.psychologytoday.com/ca/blog/animal-emotions/201506/new-zealand-declares-animals-be-sentient-bans-testing.

Bekoff, Marc. 2016a. "Why Was the Gorilla Harambe Killed at the Cincinnati Zoo?" *Psychology Today* blog, 30 May 2016. online.

Bekoff, Marc. 2016b. "Why Was Harambe the Gorilla in a Zoo in the First Place?" *Scientific American* blog, 31 May 2016. online.

Bekoff, Marc. 2018. "Why People Should Care About Animal and Human Suffering." *Psychology Today*, 1 December. Web.

Bekoff, Marc, ed. 2013. *Ignoring Nature No More*. Chicago: University of Chicago Press.

Bekoff, Marc and Jessica Pierce. 2009. *Wild Justice: The Moral Lives of Animals*. Chicago: University of Chicago Press.

Beran, Michael J. and William D. Hopkins. 2018. "Self-Control in Chimpanzees Relates to General Intelligence." *Current Biology* 28.4, 574–579e3. doi: 10.1016/j.cub.2017.12.043.

Beran, Michael J., et al. 2014. "Chimpanzees (Pan troglodytes) Can Wait, When They Choose To: A Study With the Hybrid Delay Task." *Animal Cognition* 17.2, 197–205. doi: 10.1007/s10071-013-0652-9.

Bermúdez, José Luis. 2003. *Thinking Without Words*. Oxford: OUP.

Best, Steven and Anthony J. Nocella II, eds. 2004. *Terrorists or Freedom Fighters? Reflections on the Liberation of Animals*. NY: Lantern Books.

Blackburn, Andrea, et al. 2019. "Spatial Parameters Influence the Distribution of Orangutan (*Pongo pygmaeus wurmbii*) Dispersed Seeds." National Science Foundation, in press. https://par.nsf.gov/biblio/10067612.

Bliss-Moreau, Eliza. 2017. "Constructing Nonhuman Animal Emotion." *Current Opinion in Psychology* 17, 184–188.

Bloom, Paul. 2010. "The Moral Life of Babies." *The New York Times Magazine*, 5 May. Online.

Boehm, Christopher. 2000. "Conflict and the Evolution of Social Control." *Journal of Consciousness Studies* 7.1/2, 79–101.

Boesch, Christophe, et al. 2010. "Altruism in Forest Chimpanzees: The Case of Adoption." *Plos One* 5, e8901. doi: 10.1371/journal.pone.0008901.

Boyd, Robert and Joan B. Silk. 2009. *How Humans Evolved*. Fifth edition. NY: W.W. Norton.

Brosnan, Sarah and Frans de Waal. 2003. "Monkeys Reject Unequal Pay." *Nature* 425, 297–299.

Brosnan, Sarah F. 2006. "Nonhuman Species' Reactions to Inequity and Their Implications for Fairness." *Social Justice Research* 19.2, 153–185.

Brosnan, Sarah, et al. 2005. "Tolerance for Inequity May Increase with Social Closeness in Chimpanzees." *Proceedings of the Royal Society, B* 272.1560, 253–258. doi: 10.1098/rspb.2004.2947.

Brosnan, Sarah, et al. 2011. "Orangutans (*Pongo pygmaeus*) Do Not Form Expectations Based on Their Partner's Outcomes." *Folia Primatol* 82, 56–70.

Brown, Jerram L. 1975. *The Evolution of Behavior*. NY: Norton.

Bugnyar, Thomas. 2017. "Apes Perform Like Infants in False-belief Tasks." *Learning and Behavior* 45, 325–326. doi: 10.3758/s13420-017-0268-z.

Buss, David M. 2008. *Evolutionary Psychology: The New Science of the Mind*. Third edition. Boston: Pearson.

Buttelmann, David, et al. 2017. "Great Apes Distinguish Truth from False Beliefs in an Interactive Helping Task." *Plos One* 12.4, e0173793.

Byrne, Richard W. 1995. *The Thinking Ape: Evolutionary Origins of Intelligence*. Oxford: OUP.

Byrne, Richard W. 2000. "Evolution of Primate Cognition." *Cognitive Science* 24.3, 543–570.

Byrne, Richard W., et al. 2017. "Great Ape Gestures: Intentional Communication With a Rich Set of Innate Signals." *Animal Cognition* 20, 755–769.

Calder, William A. 1984. *Size, Function, and Life History*. Cambridge, MA: Harvard UP.

Call, Josep. 2000. "Estimating and Operating on Discrete Quantities in Orangutans (*Pongo pygmaeus*)." *Journal of Comparative Psychology* 114.2, 136–147. doi: 10.1037//0735-7036.114.2.136.

Call, Josep. 2001. "Object Permanence in Orangutans (*Pongo pygmaeus*)." *Journal of Comparative Psychology* 115.2, 159–171. doi: 10.1037//0735-7036.115.2.159.

Call, Josep and Michael Tomasello. 1994. "Production and Comprehension of Referential Pointing by Orangutans (*Pongo pygmaeus*)." *Journal of Comparative Psychology* 108.4, 307–317.

Call, Josep and Michael Tomasello. 2008. "Does the Chimpanzee Have a Theory of Mind? 30 Years Later." *Trends in Cognitive Science* 12.5, 187–192.

Campbell, Liz A.D. 2019. "Fostering of a Wild, Injured, Juvenile by a Neighboring Group: Implications for Rehabilitation and Release of Barbary Macaques Confiscated From Illegal Trade." *Primates* 60, 339–345. https://doi.org/10.1007/s10329-019-00729-w.

Carne, Charlotte, et al., 2014. "The Risk of Disease to Great Apes: Simulating Disease Spread in Orang-utan (*Pongo pygmaeus wurmbii*) and Chimpanzee (*Pan troglodytes schweinfurthii*) Association Networks." *Plos One* 9.4, e95039. doi: 10.1371/journal.pone.0095039.

Caruso, Catherine. 2016. "Chimpanzees May Be Capable of Comprehending the Minds of Others." *Scientific American*, 6 October 2016. Online.

Caughlin, T. Trevor, et al. 2015. "Loss of Animal Seed Dispersal Increases Extinction Risk in a Tropical Tree Species Due to Pervasive Negative Density Dependence Across Life Stages." *Proceedings of the Royal Society B* 282, 20142095. doi: 10.1098/rspb.2014.2095.

Cavalieri, Paola. 2015. "The Meaning of the Great Ape Project." *Politics and Animals* 1, 16–34. Online.

Cavalieri, Paola. 2016. "Ethics, Animals and the Nonhuman Great Apes." *Journal of Biosciences* 31.5, 509–512.

Cavalieri, Paola and Peter Singer, eds. 1993. *The Great Ape Project: Equality Beyond Humanity*. NY: St. Martin's.

Chalmeau, Raphaël, et al. 1997. "Cooperative Problem Solving by Orangutans." *International Journal of Primatology* 18.1, 23–32.

Chang, Liangtang, et al. 2017. "Spontaneous Expression of Mirror Self-recognition in Monkeys After Learning Precise Visual-proprioceptive Association for Mirror Images." *PNAS* 114.2, 3258–3263.

Chapman, Colin A., et al. 2005. "Thirty Years of Research in Kibale National Park, Uganda, Reveals a Complex Picture for Conservation." *International Journal of Primatology* 26.3. doi: 10.1007/s10764-005-4365-z.

Chapman, Colin A., et al. 2013. "Are Primates Ecosystem Engineers?" *International Journal of Primatology* 34, 1–14. doi: 10.1007/s10764-012-9645-9.

Cheney, Dorothy L. and Robert M. Seyfarth. 1990. *How Monkeys See the World: Inside the Mind of Another Species*. Chicago: University of Chicago Press.

Chevalier-Skolnikoff, Susan, Birute Galdikas, and Alan Skolnikoff. 1982. "The Adaptive Significance of High Intelligence in Wild Orangutans: A Preliminary Report." *Journal of Human Evolution* 11, 639–652.

Clay, Zanna, et al. 2015. "Functional Flexibility in Wild Bonobo Vocal Behaviour." *PeerJ* 3, e1124. Online.

Cocks, Leif. 2016. *Orangutans: My Cousins, My Friends*. South Perth, Australia: The Orangutan Project.

Cohen, Carl. 1993. "The Case for the Use of Animals in Biomedical Research." *Moral Controversies*. Stephen Jay Gould, ed. Belmont, CA: Wadsworth, 575–582.

Conklin-Brittain, Nancy L., Cheryl D. Knott, and Richard W. Wrangham. 2001. "The Feeding Ecology of Apes." The Apes' Challenges for the 21st Century. Conference Proceedings, Brookfield Zoo. Online.

Cooper, David E., ed. 1998. *Ethics: The Classic Readings*. Oxford: Blackwell.

Coventry, Angela and Avram Hiller. 2015. "Hume on Animals and the Rest of Nature." *Animal Ethics and Philosophy: Questioning the Orthodoxy*. Elisa Aaltola and John Hadley, eds. London: Rowman and Littlefield, 165–184.

Cowdin, Daniel. 2008. "Environmental Ethics." *Theological Studies* 69, 164–184.

Cowley, Stephen and David Spurrett. 2002. "Putting Apes (Body and Language) Together Again." *Language Sciences* 25, 289–318.

Crist, Eileen. 2019. *Abundant Earth: Toward and Ecological Civilization*. Chicago: University of Chicago Press.

Crockford, Catherine, et al. 2017. "Vocalizing in Chimpanzees is Influenced by Social-Cognitive Processes." *Science Advances* 3, e1701742.

Cromsigt, Joris P. G. M., et al. 2018. "Trophic Rewilding as a Climate Change Mitigation Strategy?" *Philosophical Transactions of the Royal Society B* 373.1761. doi: 10.1098/rstb.2017.0440.

Crump, Andrew, et al. 2018. "Affect Driven Attention Biases as Animal Welfare Indicators: Review and Methods." *Animals* 8, 136. Online.

Cupp, Richard L. 2016. "Gorilla's Death Calls for Human Responsibility, Not Animal Personhood." *The Conversation*, 3 June. Online.

Curry, Oliver Scott. 2018. "Seven Moral Rules Found All Around the World." *This View of Life: Can an Evolutionary Perspective Reveal a Universal Morality*, 40–41. Online.

Czech, Brian. 2016. "Ecological Economics Comes of Age." *The Good Crisis*. John Seager and Lee S. Polansky, eds. Population Connection, 36–66.

Damasio, Antonio. 1999. *The Feeling of What Happens: Body and Emotion in the Making of Consciousness*. Orlando: Harvest/Harcourt.

Damerius, Laura, et al. 2017. "Orientation Towards Humans Predict Cognitive Performance in Orang-utans." *Scientific Reports* 7, 40052. doi: 10.1038/srep40052.

Damerius, Laura A., et al. 2019. "General Cognitive Abilities in Orangutans (*Pongo abelii* and *Pongo pygmaeus*)." *Intelligence* 74, 3–11. https://doi.org/10.1016/j.intell.2018.10.008.

Daniels, Bryan C., et al. 2016. "Quantifying Collectivity." *Current Opinion in Neurobiology* 37, 106–113. doi: 10.1016/j.conb.2016.01.012.

Darwin, Charles. 1859. *On the Origin of Species*. First edition. Joseph Carroll, ed. Canada: Broadview, 2003.

Darwin, Charles. 1871. *The Descent of Man*. Second edition. London: Penguin, 2004.

Darwin, Charles. 1872. *The Expression of the Emotions in Man and Animals*. Second edition. London: Penguin, 2009.

Dawkins, Richard. 1982. *The Extended Phenotype: The Long Reach of the Gene*. Oxford: OUP.

Dawkins, Richard. 1993. "Gaps in the Mind." *The Great Ape Project: Equality Beyond Humanity*. Paola Cavalieri and Peter Singer, eds. NY: St. Martin's, 80–87.

Dawkins, Richard. 1976. *The Selfish Gene*. Oxford: OUP, 2006.

DeCasien, Alex R., et al. 2017. "Primate Brain Size is Predicted by Diet But Not Sociality." *Nature Ecology and Evolution* 1, 112. doi: 10.1038/s41559-017-0112.

Delon, Nicolas. 2015. "Against Moral Individualism." *Animal Ethics and Philosophy: Questioning the Orthodoxy*. Elisa Aaltola and John Hadley, eds. London: Rowman and Littlefield, 31–45.

Delon, Nicolas. 2019. "Beyond the Personhood Paradigm." *ASEBL Journal* 14.1, 26–30. Online.

Desmond, Jane C. 2012. "Can Animals Make 'Art'?" *Experiencing Animal Minds*. Julie A. Smith and Robert W. Mitchell, eds. NY: Columbia University Press, 95–108.

de Waal, Frans. 1982. *Chimpanzee Politics: Power and Sex Among Apes*. Baltimore, MD: Johns Hopkins UP, 2007.

de Waal, Frans B. M. 1996a. *Good Natured: The Origin of Right and Wrong in Humans and Other Animals*. Cambridge, MA: Harvard UP.

de Waal, Frans B. M. 1996b. "Conflict as Negotiation." *Great Ape Societies*. William C. McGrew, Linda Marchant, and Toshisada Nishida, eds. Cambridge: Cambridge University Press, 159–172.

de Waal, Frans B. M. 1999. "Anthropomorphism and Anthropodenial: Consistency in our Thinking about Humans and Other Animals." *Philosophical Topics* 27.1, 255–280.

de Waal, Frans B. M. 2001. *The Ape and the Sushi Master: Cultural Reflections of a Primatologist*. NY: Basic Books.

de Waal, Frans B.M. 2006. *Primates and Philosophers: How Morality Evolved*. Princeton: Princeton UP.

de Waal, Frans B.M. 2009. *The Age of Empathy: Nature's Lessons for a Kinder Society*. NY: Harmony Books.

de Waal, Frans B.M. 2013. *The Bonobo and the Atheist: In Search of Humanism among the Primates*. NY: Norton.

de Waal, Frans B.M. 2016a. "Apes Know What Others Believe." *Science* 354.6308: 39–40.

de Waal, Frans B.M. 2016b. *Are We Smart Enough to Know How Smart Animals Are?* NY: Norton.

de Waal, Frans B.M. 2019. *Mama's Last Hug: Animal Emotions and What They Tell Us About Ourselves*. NY: Norton.

Diogo, Rui, Julia L. Molnar, and Bernard Wood. 2017. "Bonobo Anatomy Reveals Stasis and Mosaicism in Chimpanzee Evolution, and Supports Bonobos as the Most Appropriate Extant Model for the Common Ancestor of Chimpanzees and Humans." *Scientific Reports* 7, 608. doi: 10.1038/s41598-017-00548-3.

Donaldson, Sue and Will Kymlicka. 2011. *Zoopolis: A Political Theory of Animal Rights*. Oxford: OUP.

Douglas, Pamela Heidi and Liza R. Muscovice. 2015. "Pointing and Pantomime in Wild Apes? Female Bonobos Use Referential and Iconic Gestures to Request Genito-genital Rubbing." *Scientific Reports* 5, 13999. Online.

Droege, Paula. 2012. "Assessing Evidence for Animal Consciousness." *Experiencing Animal Minds*. Julie A. Smith and Robert W. Mitchell, eds. NY: Columbia University Press, 231–245.

Dubreuil, Laurent and Sue Savage-Rumbaugh. 2019. *Dialogues on the Human Ape*. Minneapolis: U Minnesota P.

Dufour, Valérie et al. 2009. "Calculated Reciprocity After All: Computation Behind Token Transfers in Orang-utans." *Biology Letters* 5, 172–175. doi: 10.1098/rsbl.2008.0644.

Dufour, Valérie et al. 2015. "Chimpanzee Drumming: A Spontaneous Performance with Characteristics of Human Musical Drumming." *Scientific Reports* 5, 11320. Online.

Dunbar, Robin I.M. 1998. "The Social Brain Hypothesis." *Evolutionary Anthropology* 6.5, 178–190.

Dunbar, R.I.M. and Susanne Shultz. 2017. "Why Are There So Many Explanations for Primate Brain Evolution?" *Philosophical Transactions of the Royal Society B* 372, 20160244.

Edwards, Paul and Arthur Pap, eds. 1973. *A Modern Introduction to Philosophy*. Third edition. NY: The Free Press.

Eisenreich, Benjamin R. and Benjamin Y. Hayden. 2018. "Cognitive Science: Persistent Apes are Intelligent Apes." *Current Biology* 28.4, R160–R162. doi: 10.1016/j.cub.2018.01.020.

Enel, Pierre, et al. 2016. "Reservoir Computing Properties of Neural Dynamics in Prefrontal Cortex." *Plos Computational Biology*. doi: 10.1371/journal.pcbi.1004967.

Estrada, Alejandro, et al. 2017. "Impending Extinction Crisis of the World's Primates: Why Primates Matter." *Science Advances* 3, e1600946. Online.

Estrada, Alejandro, et al. 2018. "Primates in Peril: The Significance of Brazil, Madagascar, Indonesia and the Democratic Republic of the Congo for Global Primate Conservation." *PeerJ* 6, e4869. doi: 10.7717/peerj.4869.

Essock-Vitale, Susan and Robert M. Seyfarth. 1987. "Intelligence and Social Cognition." *Primate Societies*. Barbara Smuts, Dorothy L. Cheney, Robert M. Seyfarth, Richard W. Wrangham, and Thomas T. Struhsaker, eds. Chicago: University of Chicago Press, 452–461.

Fahey, Eugene. 2018. State of N.Y. Court of Appeals. Motion number 2018-268. Decided 8 May 2018.

Feinberg, Todd E. and Jon M. Mallat. 2016. *The Ancient Origins of Consciousness: How the Brain Created Experience*. Cambridge, MA: MIT Press.

Fitch, W. Tecumseh, et al. 2016. "Monkey Vocal Tracts are Speech-Ready." *Science Advances* 2, e1600723. Online.

Flack, Jessica C. and Frans B.M. de Waal. 2002. "'Any Animal Whatever': Darwinian Building Blocks of Morality in Monkeys and Apes." *Evolutionary Origins of Morality: Cross-Disciplinary Perspectives*. Leonard D. Katz, ed. Thorveton, UK: Imprint Academic, 1–29.

Flemming, Timothy M., et al. 2012. "The Endowment Effect in Orangutans." *International Journal of Comparative Psychology* 25, 285–298.

Forcina, Giovanni, et al. 2019. "From Groups to Communities in Western Lowland Gorillas." *Proceedings of the Royal Society B* 286, 20182019. doi: 10/1098/rspb.2018.2019.

Foreman, Elizabeth. 2015. "Doing Without Moral Rights." *Animal Ethics and Philosophy: Questioning the Orthodoxy*. Elisa Aaltola and John Hadley, eds. London: Rowman and Littlefield, 133–147.

Fossey, Dian. 1983. *Gorillas in the Mist*. Boston: Houghton Mifflin, 2000.

Fouts, Roger S. 1972. "The Use of Guidance in Teaching Sign Language to a Chimpanzee." *Journal of Comparative and Physiological Psychology* 80, 515–522.

Fouts, Roger S. 1973. "Acquisition and the Testing of Gestural Signs in Four Young Chimpanzees." *Science* 180, 978–980.

Fouts, Roger S. and Deborah H. Fouts. 1993. "Chimpanzees' Use of Sign Language." *The Great Ape Project: Equality Beyond Humanity*. Paola Cavalieri and Peter Singer, eds. NY: St. Martin's, 28–41.

Fox, Elizabeth A. 2002. "Female Tactics to Reduce Sexual Harassment in the Sumatran Orangutan (*Pongo pygmaeus abelii*)." *Behavioral Ecology and Sociobiology* 52.2, 93–101.

Fox, Elizabeth A., Arnold F. Sitompul, and Carel van Schaik. 1999. "Intelligent Toold Use in Wild Sumatran Orangutans." *The Mentalities of Gorillas and Orangutans: Comparative Perspectives*. Sue Taylor Parker, Robert W. Mitchell, and H. Lyn Miles, eds. Cambridge: Cambridge University Press, 99–116.

Fox, Justin. 2012. "Great Ape Personhood." *Law School Student Scholarship* 102. https://scholarship.shu.edu/student_scholarship/102.

Frölich, Marlen, et al. 2019. "The Loud Scratch: A Newly Identified Gesture of Sumatran Orangutan Mothers in the Wild." *Biology Letters* 15, 20190209. doi: 10.1098/rsbl.2019.0209.

Fruth, Barbara and Gottfried Hohmann. 1996. "Nest Building Behavior in the Great Apes: The Great Leap Forward?" *Great Ape Societies*. William C. McGrew, Linda Marchant, and Toshisada Nishida, eds. Cambridge: Cambridge University Press, 225–240.

Fruth, Barbara and Gottfried Hohmann. 2018. "Food Sharing Across Borders: First Observation of Intercommunity Meat Sharing by Bonobos at LuiKotale, DRC." *Human Nature*. https://doi.org/10.1007/s12110-018-9311-9.

Gabora, Liane and A. Russon. 2011. "The Evolution of Intelligence." *The Cambridge Handbook of Intelligence*. R. Sternberg and S. Kaufman, eds. Cambridge: Cambridge University Press, 328–350.

Galdikas, Biruté. 1995a. "Social and Reproductive Behavior of Wild Adolescent Female Orangutans." *The Neglected Ape*. Ronald D. Naler, et al., eds. NY: Plenum, 163–182.

Galdikas, Biruté. 1978. *Orangutan Adaptation at Tanjung Puting Reserve, Central Borneo*. Ann Arbor: University Microfilms Intl.

Galdikas, Biruté and Gary Shapiro. 1996. "Orangutan Ethics." *Etica & Animali* 8, 50–67.

Galdikas, Biruté, Gary Shapiro, and Flora Katz. 1985. "Danau Burung, a Bird Lake in Southern Indonesian Borneo." *Ardea* 73, 189–190.

Galdikas, Biruté M. F. 1989. "Orangutan Tool Use." *Science* 243.4888, 152.

Galdikas, Biruté M. F. 1995b. *Reflections of Eden: My Years with the Orangutans of Borneo*. Boston: Little Brown.

Galdikas, Biruté M. F. and Stephen J. Insley. 1988. "The Fast Call of the Adult Male Orangutan." *Journal of Mammalogy* 69.2, 371–375.

Gallagher, Shaun. 2009. "Philosophical Antecedents to Situated Cognition." *The Cambridge Handbook of Situated Cognition*. P. Robbins and M. Aydede, eds. Cambridge: Cambridge University Press, 35–51.

Gallagher, Shaun. 2014a. "Phenomenology and Embodied Cognition." *Routledge Handbook of Embodied Cognition*. Lawrence Shapiro, ed. London: Routledge, 9–18.

Gallagher, Shaun. 2014b. "Pragmatic Interventions into Enactive and Extended Conceptions of Cognition." *Philosophical Issues* 24, 110–126.

Gallagher, Shaun and Robb Lindgren. 2015. "Enactive Metaphors: Learning Through Full-body Engagement." *Educational Psychology Review* 27, 391–404. doi: 10.1007/s10648-015-9327-1.

Gallup, Gordon G., Jr. 1970. "Chimpanzees: Self-Recognition." *Science* 167, 86–87.

Gallup, Gordon G., Jr. 1985. "Do Minds Exist in Species Other Than Our Own?" *Neuroscience & Biobehavioral Reviews* 9.4, 631–641.

Gardner, Andy. 2017. "The Purpose of Adaptation." *Interface Focus* 7, 2017005. Online.

Gardner, Howard. 2011. *Frames of Mind: The Theory of Multiple Intelligences*. Third edition. NY: Basic Books.

Gardner, R. Allen and Beatrix Gardner. 1969. "Teaching Sign Language to a Chimpanzee." *Science* 165, 664–672.

Garriga, Rosa M., et al. 2019. "Factors Influencing Wild Chimpanzee (*Pan troglodytes verus*) Relative Abundance in an Agricultural-swamp Matrix Outside Protected Areas." *Plos One* 14.5, e0215545. https://doi.org/10.1371/journal.pone.0215545.

Gergely, György. 1994. "From Self-recognition to Theory of Mind." *Self-Awareness in Animals and Humans: Developmental Perspectives*. Sue Taylor Parker, Robert W. Mitchell, and Maria L. Boccia, eds. Cambridge: Cambridge University Press, 51–60.

Gharajehdaghipour, Tazarve, et al. 2016. "Arctic Foxes as Ecosystem Engineers: Increased Soil Nutrients Lead to Increased Plant Production on Fox Dens." *Scientific Reports* 6, 24020. doi: 10.1038/srep24020.

Golonka, Sabrina and Andrew D. Wilson. 2012. "Gibson's Ecological Approach." *Avant* 3.2, 40–53.

Gómez, Juan Carlos. 1990. "The Emergence of Intentional Communication as a Problem-Solving Strategy in the Gorilla." *"Language" and Intelligence in Monkeys and Apes*. Sue Taylor Parker and Kathleen Rita Gibson, eds. Cambridge: Cambridge University Press, 333–355.

Gómez, Juan Carlos. 1996. "Ostensive Behavior in Great Apes: The Role of Eye Contact." *Reaching Into Thought: The Minds of the Great Apes*. Anne E. Russon, Kim A. Bard, and Sue Taylor Parker, eds. Cambridge: Cambridge University Press, 131–151.

Gonseth, Chloe, et al. 2017. "The Higher the Farther: Distance-specific Referential Gestures in Chimpanzees (*Pan troglodytes*)." *Biology Letters* 13, 20170398. Online.

Gonzales, Lauren A., et al. 2015. "Cerebral Complexity Preceded Enlarged Brain Size and Reduced Olfactory Bulbs in Old World Monkeys." *Nature Communications* 6, 7580. Online.

Goodall, Jane. 1971. *In the Shadow of Man*. Boston: Mariner.

Goodall, Jane. 1986. *The Chimpanzees of Gombe: Patterns of Behavior*. Cambridge, MA: Harvard UP.

Goodall, Jane. 1990. *Through a Window: My Thirty Years with the Chimpanzees of Gombe*. Boston: Mariner.

Goodall, Jane. 1993. "Chimpanzees – Bridging the Gap." *The Great Ape Project: Equality Beyond Humanity*. Paola Cavalieri and Peter Singer, eds. NY: St. Martin's, 10–18.

Goodin, Robert E., et al. 1997. "Simian Sovereignty." *Political Theory* 25.6, 821–849.

Goudge, Thomas Anderson. 1967. "Jakob Johann, Baron von Uexküll." *The Encyclopedia of Philosophy*. Volume 8. Paul Edwards, ed. NY: Macmillan, 173–174.

Graham, Kirsty E., et al. 2018. "Bonobo and Chimpanzee Gestures Overlap Extensively in Meaning." *Plos Biology* 16.2, e2004825. https://doi.org/10.1371/journal.pbio.2004825.

Grant, Verne. 1963. *The Origin of Adaptations*. NY: Columbia University Press.

Graziano, Michael S.A. and Taylor W. Webb. 2014. "A Mechanistic Theory of Consciousness." *International Journal of Machine Consciousness* 6.2, 1–14. doi: 10.1142/S1793843014001316.

Greenfield, Patricia Marks and E. Sue Savage-Rumbaugh. 1990. "Grammatical Combination in *Pan paniscus*: Processes of Learning and Invention in the Evolution and Development of Language." *"Language" and Intelligence in Monkeys and Apes*. Sue Taylor Parker and Kathleen Rita Gibson, eds. Cambridge: Cambridge University Press, 540–578.

Griggs, David, et al. 2013. "Sustainable Development Goals for People and Planet." *Nature* 495, 305–307.

Gruber, Thibaud, et al. 2009. "Wild Chimpanzees Rely on Cultural Knowledge to Solve an Experimental Honey Acquisition Task." *Current Biology* 19, 1806–1810.

Gruber, Thibaud, et al. 2019. "Cultural Change in Animals: A Flexible Behavioral Adaptation to Human Disturbance." *Palgrave Communications* 5, 64. https://doi.org/10.1057/s41599-019-0271-4.

Gruen, Lori. 2013. "The Moral Status of Animals." *Reflecting on Nature*, second edition. Lori Gruen, Dale Jamieson, and Christopher Schlottmann, eds. NY: OUP, 151–160.

Gruen, Lori. 2019. Foreword. *Chimpanzee Rights*. Oxford: Routledge, xiii–xix.

Grueneisen, Sebastian, et al. 2017. "Children, Chimpanzees, and Bonobos Adjust the Visibility of their Actions for Cooperators and Competitors." *Scientific Reports* 7, 8504. doi: 10.1038/s41598-017-08435-7.

Hadley, John. 2015. "A Metalevel Problem for Animal Rights Theory." *Animal Ethics and Philosophy: Questioning the Orthodoxy*. Elisa Aaltola and John Hadley, eds. London: Rowman and Littlefield, 15–30.

Haemig, P. D. 2012. "Ecosystem Engineers: Wildlife That Create, Modify and Maintain Habitats." *Ecology Info* 12. Online.

Harari, Yuval Noah. 2017. *Homo Deus: A Brief History of Tomorrow*. NY: Perennial.

Hare, Brian. 2001. "Can Competitive Paradigms Increase the Validity of Experiments on Primate Social Cognition?" *Animal Cognition* 4, 269–280.

Hare, Brian, et al. 2001. "Do Chimpanzees Know What Conspecifics Know?" *Animal Behaviour* 61, 139–151.

Harlow, Harry. 1958. "The Nature of Love." *American Psychologist* 13, 573–685.

Harlow, Harry F., et al. 1965. "Total Social Isolation in Monkeys." *PNAS* 54, 90–97.

Harrisson, Barbara. 1962. *Orang-Utan*. Oxford: OUP, 1987.

Harrison, Robert Pogue. 2008. *Gardens: An Essay on the Human Condition*. Chicago: University of Chicago Press.

Hart, Daniel and Mary Pat Karmel. 1996. "Self-awareness and Self-knowledge in Humans, Apes, and Monkeys." *Reaching Into Thought: The Minds of the Great Apes*. Anne E. Russon, Kim A. Bard, and Sue Taylor Parker, eds. Cambridge: Cambridge University Press, 325–347.

Hayek, Matthew N. and Rachael D. Garrett. 2018. "Nationwide Shift to Grass-fed Beef Requires Larger Cattle Population." *Environmental Research Letters* 13, 084005. https://doi.org/10.1088/1748-9326/aad401.

Heesen, Raphaela, et al. 2019. "Linguistic Laws in Chimpanzee Gestural Communication." *Proceedings of the Royal Society B* 286, 20182900. doi: 10.1098/rspb.2018.2900.

Heilbroner, Robert L. 1961. *The Worldly Philosophers*. Revised edition. NY: Simon and Schuster.

Hess, Elizabeth. 2008. *Nim Chimpsky: The Chimp Who Would Be Human*. NY: Bantam.

Hewes, Gordon W. 1973. "Primate Communication and the Gestural Origins of Language." *Current Anthropology* 14, 5–24.

Heyes, Cecilia. 2003. "Four Routes of Cognitive Evolution." *Psychological Review* 110.4, 713–727.

Heyes, Cecilia. 2016. "Apes Submentalise." *Trends in Cognitive Science* 21.1, 1–2.

Heyes, Cecilia M. 1998. "Theory of Mind in Nonhuman Primates." *Behavioral and Brain Sciences* 21.1, 101–114.

Hicks, Thurston C., et al. 2019. "Bili-Uéré: A Chimpanzee Behavioural Realm in Northern Democratic Republic of Congo." *Folia Primatologica* 90, 3–64. doi: 10.1159/000492998.

Hinde, Robert A. "Can Nonhuman Primates Help Us Understand Human Behaviors?" *Primate Societies*. Barbara Smuts, Dorothy L. Cheney, Robert M. Seyfarth, Richard W. Wrangham, and Thomas T. Struhsaker, eds. Chicago: University of Chicago Press, 413–420.

Hinman, Lawrence M. 2003. *Ethics: A Pluralistic Approach to Moral Theory*. Third edition. Belmont, CA: Wadsworth.

Hobaiter, Catherine, et al. 2014. "Social Network Analysis Shows Direct Evidence for Social Transmission of Tool Use in Wild Chimpanzees." *Plos Biology* 12.9. Online.

Hobbes, Thomas. 1651. *Leviathan, or Matter, Form, and Power of a Commonwealth Ecclesiastical and Civil*. Nelle Fuller, ed. Chicago: Great Books, 1952.

Hoek, Tim A., et al. 2016. "Resource Availability Modulates the Cooperative and Competitive Nature of a Microbial Cross-feeding Mutualism." *Plos Biology* 14.8, e1002540. doi: 10.1371/journal.pbio.1002540.

Hoff, Christina. 1993. "Immoral and Moral Uses of Animals." *Moral Controversies*. Stephen Jay Gould, ed. Belmont, CA: Wadsworth, 582–586.

Hohmann, Gottfried, et al. 2019. "Fishing For Iodine: What Aquatic Foraging by Bonobos Tells Us About Human Evolution." *BMC Zoology* 4, 5. https://doi.org/10.1186/s40850-019-0043-z.

Holmes, Robert L. 1993. *Basic Moral Philosophy*. Belmont, CA: Wadsworth.

Hopkins, William D., et al. 2019. "More Intelligent Chimpanzees (Pan troglodytes) Have Larger Brains and Increased Cortical Thickness." *Intelligence* 74, 18–24. https://doi.org/10.1016/j.intell.2018.11.002.

Horner, Victoria and Frans B.M. de Waal. 2009. "Controlled Studies of Chimpanzee Cultural Transmission." *Cultural Neuroscience: Cultural Influences on Brain Function*. Joan Chiao, ed. Amsterdam: Elsevier, 3–15.

Horner, Victoria, et al. 2010. "Prestige Affects Cultural Learning in Chimpanzees." *Plos One* 5.5, e10625. doi: 10.1371/journal.pone.0010625.

Horner, Victoria, et al. 2011. "Spontaneous Prosocial Choice by Chimpanzees." *PNAS* 108.33, 13847–13851.

Horstmann, Aike C., et al. 2018. "Do a Robot's Social Skills and Its Objection Discourage Interactants From Switching the Robot Off?" *Plos One* 13.7, e0201581. https://doi.org/10.1371/journal.pone.0201581.

Houghton, Walter E. 1957. *The Victorian Frame of Mind*. New Haven: Yale UP.

Hrdy, Sarah Blaffer. 2001. "Mothers and Others." *Natural History*. Online.

Hrdy, Sarah Blaffer. 2006. "Evolutionary Context of Human Development: The Cooperative Breeding Model." *Attachment and Bonding: A New Synthesis*. C. Sue Carter, Lieselotte Ahnert, K. E. Grossman, Sarah B. Hrdy, Michael E. Lamb, Stephen W. Porges, and Norbert Sachser, eds. Cambridge, MA: MIT P.

Hublin, Jean-Jacques, et al. 2017. "New Fossils from Jebel Irhoud, Morocco and the Pan-African Origin of *Homo Sapiens*." *Nature* 546, 289–292.

Hufendiek, Rebekka. 2018. *Embodied Emotions: A Naturalist Approach to a Normative Phenomenon*. NY: Routledge.

Hunt, Kevin D. 2004. "The Special Demands of Great Ape Locomotion and Posture." *The Evolution of Thought: Evolutionary Origins of Great Ape Intelligence*. Anne E. Russon and David R. Begun, eds. Cambridge: Cambridge University Press, 172–189.

Hunter, Lori M. 2016. "Environmental Dimensions of Population Stabilization." *The Good Crisis*. John Seager and Lee S. Polansky, eds. Population Connection, 155–161.

Hutchins, Edwin. 2008. "The Role of Cultural Practices in the Emergence of Modern Human Intelligence." *Philosophical Transactions of the Royal Society B* 363, 2011–2019. doi: 10.1098/rstb.2008.0003.

Jablonka, Eva and Marion J. Lamb. 2005. *Evolution in Four Dimensions: Genetic, Epigenetic, Behavioral, and Symbolic Variation in the History of Life*. Cambridge, MA: MIT/Bradford.

Jaffe, Barbara. 2015. Supreme Court of the State of New York, New York County, Part 12. The Nonhuman Rights Project, on behalf of Hercules and Leo, against, Samuel L. Stanely, et al. Index number, 152736/2015.

Jamieson, Dale. 2013. "Animal Liberation is an Environmental Ethic." *Reflecting on Nature*, second edition. Lori Gruen, Dale Jamieson, and Christopher Schlottmann, eds. NY: Oxford University Press, 167–176.

Jones, Clive G., et al. 1994. "Organisms as Ecosystem Engineers." *Oikos* 69.3, 373–386.

Jones, Clive G., et al. 1997. "Positive and Negative Effects of Organisms as Physical Ecosystem Engineers." *Ecology* 78, 1946–1957.

Kano, Fumihiro, et al. 2017. "Submentalizing Cannot Explain Belief-Based Action Anticipation in Apes." *Trends in Cognitive Science* 21.9, 633–634.

Kano, Fumihiro, et al. 2019. "Great Apes Use Self-experience to Anticipate an Agent's Action in a False Belief Test." *PNAS*. https://doi.org/10.1073/pnas.1910095116.

Kaplan, Gisela and Lesley J. Rogers. 2000. *The Orangutans: Their Evolution, Behavior, and Future*. Cambridge, MA: Perseus.

Kappeler, Peter M., Joan B. Silk, Judith Burkart, and Carel P. van Schaik. 2010. "Primate Behavior and Human Universals: Exploring the Gap." *Mind the Gap: Tracing the Origins of Human Universals*. Peter M. Kappeler and Joan B. Silk, eds. Heidelberg: Springer, 3–15.

King, Barbara J. 2004. *The Dynamic Dance: Nonvocal Communication in African Great Apes*. Cambridge, MA: Harvard UP.

King, Barbara J. 2013. *We All Feel: Understanding Animal Grief and Love*. Chicago: University of Chicago Press Shorts. E-book.

King, Barbara J. 2017. *Personalities on the Plate: The Lives and Minds of Animals We Eat*. Chicago: University of Chicago Press.

Kittay, Eva Feder. 2005. "At the Margins of Moral Personhood." *Ethics* 116.1, 100–131.

Knott, Cheryl D. 1998. "Changes in Orangutan Caloric Intake, Energy, Balance, and Ketones in Response to Fluctuating Fruit Availability." *International Journal of Primatology* 19.6, 1061–1079.

Knott, Cheryl, et al. 2008. "Female-Female Competition in Bornean Orangutans." *International Journal of Primatology* 29, 975–997. doi: 10.1007/s10764-008-9278-1.

Koel, Todd M., et al. 2019. "Predatory Fish Invasion Induces Within and Across Ecosystem Effects in Yellowstone National Park." *Science Advances* 5, eaav1139. Online.

Köhler, Wolfgang. 1925. *The Mentality of Apes*. Ella Winter, trans. London: Routledge and Kegan Paul, 1973.

Koops, Kathelijne, et al. 2014. "The Ecology of Primate Material Culture." *Biology Letters* 10. Online.

Koops, Kathelijne, et al. 2015a. "Chimpanzees and Bonobos Differ in Intrinsic Motivation for Tool Use." *Scientific Reports* 5, 11356.

Koops, Kathelijne, et al. 2015b. "Cultural Differences in Ant-dipping Tool Length Between Neighbouring Chimpanzee Communities at Kalinzu, Ugana." *Scientific Reports* 5, 12456.

Korb, Margaret P., et al. 1989. *Gestalt Therapy: Practice and Theory*. Second edition. NY: Pergamon P.

Korsgaard, Christine M. 2013. "Personhood, Animals, and the Law." *Think* 12.34, 25–32. Online.

Korsgaard, Christine M. 2018. *Fellow Creatures: Our Obligations to Other Animals*. Oxford: OUP.

Kret, Mariska E. and Masaki Tomonaga. 2016. "Getting to the Bottom of Face Processing: Species-Specific Inversion Effects for Faces and Behinds in Humans and Chimpanzees (*Pan Troglodytes*)." *Plos One* 11.11, e0165357. doi: 10.1371/journal.pone.0165357.

Kret, Mariska E., et al. 2016. "Bonobos (*Pan paniscus*) Show an Attentional Bias Toward Conspecifics' Emotions." *PNAS Early Edition*. Online.

Kropotkin, Peter. 1902. *Mutual Aid: A Factor of Evolution*. NY: McClure Phillips.

Krupenye, Christopher, et al. 2016. "Great Apes Anticipate that Other Individuals Will Act According to False Beliefs." *Science* 354.6308, 110–114.

Kühl, Hjalmar S., et al. 2016. "Chimpanzee Accumulative Stone Throwing." *Scientific Reports* 6, 22219.

Lambert, Joanna. 1998. "Primate Frugivory in Kibale National Park, Uganda, and Its Implications for Human Use of Forest Resources." *African Journal of Ecology*. https://doi.org/10.1046/j.1365-2028.1998.00131.x.

Lambert, Joanna and Paula Garber. 1998. "Evolutionary and Ecological Implications of Primate Seed Dispersal." *American Journal of Primatology* 45, 9–28.

Lameira, Adriano R. and Josep Call. 2018. "Time-Space-Displaced Responses in the Orangutan Vocal System." *Science Advances* 4, eaau3401.

Lameira, Adriano R., et al. 2016. "Vocal Fold Control Beyond the Species-Specific Repertoire in an Orang-utan." *Scientific Reports* 6, 30315. doi: 10.1038/srep30315.

Latzman, Robert D., et al. 2014. "Personality in Chimpanzees (*Pan troglodytes*): Exploring the Hierarchical Structure and Associations with the Vasopressin V1A Receptor Gene." *Plos One* 9.4, e95741. doi: 10.1371/journal.pone.0095741.

Latzman, Robert D., et al. 2015. "Neuroanatomical Correlates of Personality in Chimpanzees (*Pan troglodytes*): Associations Between Personality and Frontal Cortex." *NeuroImage* 123, 63–71. Online.

Laumer, Isabelle B., et al. 2018. "Spontaneous Innovation of Hook-bending and Unbending in Orangutans (*Pongo abelii*)." *Scientific Reports* 8, 16518. doi: 10.1038/s41598-018-34607-0.

Laumer, Isabelle B., et al. 2019. "Orangutans (*Pongo abelii*) Make Flexible Decisions Relative to Reward Quality and Tool Functionality in a Multi-dimensional Tool-use Task." *Plos One* 14.2, e0211031. https://doi.org/10.1371/journal.pone.0211031.

Leavens, David A., et al. 2017. "The Mismeasure of Ape Social Cognition." *Animal Cognition*. doi: 10.1007/s10071-017-1119-1.

Lee, A.T.K., et al. 2019. *Reforesting for the Climate of Tomorrow*. Gland, Switzerland: IUCN.

Leitan, Nuwan D. and Lucian Chaffey. n.d. "Embodied Cognition and Its Applications." *Sensoria: A Journal of Mind, Brain and Culture*. Online.

Leitch, Carmen. 2019. "Controversial Work Adds Human Brain Gene to Rhesus Monkey Genome." *Labroots*, 15 April. Online.

Leopold, Aldo. 1949. *A Sand County Almanac*. Oxford: OUP.

Lethmate, Jurgens and Gerti Ducker. 1973."Experiments on Self-Recognition in a Mirror in Orangutans, Chimpanzees, Gibbons and Several Monkey Species." *Zeitschrift für Tierpsychologie* 33.3–4, 248–269.

Lewin, Roger. 1983. "Is the Orangutan a Living Fossil?" *Science* 222.4629, 1222–1223.

Li, Xiang, et al. 2017. "Human Torque is Not Present in Chimpanzee Brain." *Neuroimage* 165, 285–293.

Linzey, Andrew. 2009. *Why Animal Suffering Matters*. Oxford: OUP.

Lorenz, Konrad. 1965. *Evolution and Modification of Behavior*. Chicago: University of Chicago Press.

Lotem, Arnon, et al. 2017. "The Evolution of Cognitive Mechanisms in Response to Cultural Innovations." *PNAS* 114.30, 7915–7922.

Lovelock, James. 1979. *Gaia: A New Look at Life on Earth*. Oxford: OUP, 2000.

Luncz, Lydia V., et al. 2018. "Costly Culture: Differences in Nut-cracking Efficiency Between Wild Chimpanzee Groups." *Animal Behaviour* 137, 63–73.

Lurz, Robert W. 2012. "Chimpanzees Attribute Beliefs?: A New Approach to Answering and Old Nettled Question." *Experiencing Animal Minds*. Julie A. Smith and Robert W. Mitchell, eds. NY: Columbia University Press, 263–285.

Lustgarten, Abrahm. 2018. "Palm Oil Was Supposed to Help Save the Planet. Instead It Unleased a Catastrophe." *New York Times Magazine*, 20 November. Online.

Lwanga, J. S., et al. 2011. "Primate Population Dynamics Over 32.9 Years at Ngogo, Kibale National Park, Uganda." *American Journal of Primatology* 73, 997–1011.

MacLean, Evan L., et al. 2014. "The Evolution of Self-control." *PNAS*, E2140–E2148. www.pnas.org/cgi/doi/10.1073/pnas.1323533111.

Maggioncalda, Anne, Robert Sapolsky and Nancy Czekala. 1999. "Reproductive Hormone Profiles in Captive Male Orangutans: Implications for Understanding Developmental Arrest." *American Journal of Physical Anthropology* 109.1, 19–32.

Malle, Bertram F. and Jess Scon Holbrook. 2009. "Theory of Mind and Consciousness." *The Oxford Companion to Consciousness*. Tim Bayne, Alex Cleeremans, and Patrick Wilken, eds. Oxford: OUP, 630–632.

Malone, Nicholas, et al. 2014. "Ethnoprimatology: Critical Interdisciplinarity and Multispecies Approaches in Anthropology." *Critique of Anthropology* 34.1, 8–29.

Markham, A. Catherine and Laurence R. Gesquiere. 2016. "Costs and Benefits of Group Living in Primates: An Energetic Perspective." *Transactions of the Royal Society B* 372, 20160239. Online.

Marsh, Heidi L. 2012. "Orangutans' Use of Contiguous Versus Distal Social and Non-social Cues in an Object-choice Task." *International Journal of Comparative Psychology* 25, 299–308.

Marshall-Pescini, Sarah, and Andrew Whiten. 2008. "Chimpanzees (Pan trogloytes) and the Question of Cumulative Culture: An Experimental Approach." *Animal Cognition* 11, 449–456.

Martin, Alia and Laurie R. Santos. 2016. "What Cognitive Representations Support Primate Theory of Mind?" *Trends in Cognitive Science* 20.5, 375–382.

Masserman, Jules H., Stanley Wechkin, and William Terris. 1964. "'Altruistic' Behavior in Rhesus Monkeys." *American Journal of Psychiatry* 121, 584–585.

Mayer, Emeran A. 2011. "Gut Feelings: The Emerging Biology of Gut-Brain Communication." *Nature Reviews* 12: 453–466.

McCoy, Dakota E., et al. 2019. "New Caledonian Crows Behave Optimistically After Using Tools." *Current Biology* 29, 1–6. https://doi.org/10.1016/j.cub.2019.06.080.

McGrew, W.C. 1998. "Culture in Nonhuman Primates?" *Annual Review of Anthropology* 27, 301–328.

McGrew, William C. 2010. "Chimpanzee Technology." *Science* 328, 579–580.

McGrew, William C., Linda Marchant, and Toshisada Nishida, eds. 1996. *Great Ape Societies*. Cambridge: Cambridge University Press.

McMahan, Jeff. 2005. "Our Fellow Creatures." *The Journal of Ethics* 9, 353–380. doi: 10.1007/s10892-005-3512-2.

McNeely, Jeffrey and Paul Spencer Wachtel. 1988. *Soul of the Tiger*. NY: Doubleday.

Meijer, Eva. 2019. *When Animals Speak: Toward an Interspecies Democracy*. NY: NYU Press.

Mendl, Michael and Elizabeth S. Paul. 2008. "Do Animals Live in the Present?: Current Evidence and Implications for Welfare." *Applied Animal Behavior Science* 113.4, 357–382. https://doi.org/10.1016/j.applanim.2008.01.013.

Menzel, Jr., E.W. 1971. "Communication About the Environment in a Group of Young Chimpanzees." *Folia Prima Primatologica* 15, 220–232.

Mercader, Julio, et al. 2007. "4,300-Year Old Chimpanzee Sites and the Origins of Percussive Stone Technology." *PNAS* 104.9, 3043–3048.

Mesoudi, Alex. 2011. *Cultural Evolution: How Darwinian Theory Can Explain Human Culture and Synthesize the Social Sciences*. Chicago: University of Chicago Press.

Michaels, Claire F. and Zsolt Palatinus. 2014. "A Ten Commandments for Ecological Psychology." *The Routledge Handbook of Embodied Cognition*. Lawrence Shapiro, ed. London: Routledge.

Miles, H. Lyn. 1999. *The Mentalities of Gorillas and Orangutans: Comparative Perspectives*. Sue Taylor Parker, Robert W. Mitchell, and H. Lyn Miles, eds. Cambridge: Cambridge University Press, 197–210.

Miles, H. Lyn White. 1990. "The Cognitive Foundations for Reference in a Signing Orangutan." *"Language" and Intelligence in Monkeys and Apes*. Sue Taylor Parker and Kathleen Rita Gibson, eds. Cambridge: Cambridge University Press, 511–539.

Miles, H. Lyn White. 1993. "Language and the Orang-utan: The Old 'Person' of the Forest." *The Great Ape Project: Equality Beyond Humanity*. Paola Cavalieri and Peter Singer, eds. NY: St. Martin's, 42–57.

Miles, H. Lyn White. 1994. "ME CHANTEK: The Development of Self-awareness in a Signing Orangutan." *Self-Awareness in Animals and Humans: Developmental Perspectives*. Sue Taylor Parker, Robert W. Mitchell, and Maria L. Boccia, eds. Cambridge: Cambridge University Press, 254–272.

Miles, H. Lyn, Robert W. Mitchell, and Stephen E. Harper. 1996. "Simon Says: The Development of Imitation in an Encultured Orangutan." *Reaching Into Thought: The Minds of the Great Apes*. Anne E. Russon, Kim A. Bard, and Sue Taylor Parker, eds. Cambridge: Cambridge University Press, 278–299.

Mitchell, Matthew and Mary Katherine Gonder. 2013. "Primate Speciation: A Case Study of African Apes." *Nature Education Knowledge* 4.2. Online.

Mitchell, Robert W. 1993. "Humans, Nonhumans and Personhood." *The Great Ape Project: Equality Beyond Humanity*. Paola Cavalieri and Peter Singer, eds. NY: St. Martin's, 237–247.

Mitchell, Robert W. 1994. "Multiplicities of Self." *Self-awareness in Animals and Humans: Developmental Perspectives*. Sue Taylor Parker, Robert W. Mitchell, and Maria L. Boccia, eds. Cambridge: Cambridge University Press, 81–107.

Mitchell, Robert. 1997. "Kinesthetic-Visual Matching and the Self-Concept as Explanations of Mirror-Self-Recognition." *Journal for the Theory of Social Behavior* 27, 101–123.

Mitchell, Robert W. 1999. "Deception and Concealment as Strategic Script Violation in Great Apes and Humans." *The Mentalities of Gorillas and Orangutans: Comparative Perspectives*. Sue Taylor Parker, Robert W. Mitchell, and H. Lyn Miles, eds. Cambridge: Cambridge University Press, 295–315.

Moffett, Mark W. 2019. *The Human Swarm: How Our Societies Arise, Thrive, and Fall*. NY: Basic Books.

Moore, Richard. 2016. "Meaning and Ostension in Great Ape Gestural Communication." *Animal Cognition* 19.1, 223–231.

Morell, Virginia. 2016. "Humans Aren't the Only Great Apes That Can 'Read Minds.'" *Science*, 6 October 2016. Online.

Morin, Alain. 2012. "What are Animals Conscious Of?" *Experiencing Animal Minds*. Julie A. Smith and Robert W. Mitchell, eds. NY: Columbia University Press, 246–260.

Morrison, Robin E., et al. 2019. "Hierarchical Social Modularity in Gorillas." *Proceedings of the Royal Society B* 286, 20190681. doi: 10.1098.rspb.2019.0681.

Morrogh-Bernard, Helen C., et al. 2008. "Orangutan Activity Budgets and Diet: A Comparison Between Species, Populations and Habitats." *Orangutans: Geographic Variation in Behavioral Ecology and Conservation*. Serge A. Wich, ed. Oxford: OUP, 119–133.

Morrogh-Bernard, H. C., et al. 2017. "Self-medication by Orang-Utans (*Pongo pygmaeus*) Using Bioactive Properties of *Dracaena Cantleyi*." *Scientific Reports* 7, 16653. doi: 10.1038/s41598-017-16621-w.

Moses, Louis J. 1994. "Foreword." *Self-awareness in Animals and Humans: Developmental Perspectives*. Sue Taylor Parker, Robert W. Mitchell, and Maria L. Boccia, eds. Cambridge: Cambridge University Press, x–xvi.

Mosley, Hannah and Michael Haslam. 2017. "Extending Material Cognition to Primate Tool Use." *Quaternary International* 405A, 70–77.

Motes-Rodrigo, Alba, et al. 2019. "Chimpanzee Extractive Foraging with Excavating Tools: Experimental Modeling of the Origins of Human Technology." *Plos One* 14.5, e0215644. https://doi.org/10.1371/journal.pone.0215644.

Mulcahy, Nicholas J. 2016. "Orangutans (*Pongo abelii*) Seek Information About Tool Functionality in a Metacognition Tubes Task." *Journal of Comparative Psychology* 130.4, 391–399. doi: 10.1037/com0000046.

Mulcahy, Nicholas J. 2018. "An Orangutan Hangs Up a Tool For Future Use." *Scientific Reports* 8, 12900. doi: 10.1038/s41598-018-31331-7.

Mulcahy, Nicholas J. and Josep Call. 2006. "Apes Save Tools for Future Use." *Science* 312, 1038–1040.

Mulcahy, Nicholas J. and Josep Call. 2009. "The Performance of Bonobos (*Pan paniscus*), Chimpanzees (*Pan troglodytes*), and Orangutans (*Pongo pygmaeus*) in Two Versions of an Object-Choice Task." *Journal of Comparative Psychology* 123.3, 304–309. doi: 10.1037/a0016222.

Musgrave, Stephanie, et al. 2016. "Tool Transfers are a Form of Teaching Among Chimpanzees." *Scientific Reports* 6, 34783. doi: 10.1038/srep34783.

Nadler, Ronald D., et al., eds. 1995. *The Neglected Ape*. NY: Plenum.

Nagel, Thomas. 1979. "What Is It Like to Be a Bat?" *Mortal Questions*. London: Cambridge UP, 165–180.

Nagel, Thomas. 2019. "What We Owe a Rabbit." Review of Christine Korsgaard's *Fellow Creatures*. *New York Review of Books*, 21 March. Online.

Nater, Alexander, et al. 2017. "Morphometric, Behavioral, and Genomic Evidence for a New Orangutan Species." *Current Biology* 27, 3487–3498.

Navarette, Ana F., et al. 2016. "The Coevolution of Innovation and Technical Intelligence in Primates." *Philosophical Transactions of the Royal Society B* 371, 20150186. Online.

Nonhuman Rights Project v. Patrick C. Lavery et al. 2018. Amicus Curiae brief. New York State Court of Appeals. NY County Clerk's Index Numbers: 162358/15 and 150149/16.

Norman, Richard. 1998. *The Moral Philosophers: An Introduction to Ethics*. Second edition. Oxford: OUP.

Nosworthy, Jane. 1998. "The Koko Dilemma: A Challenge to Legal Personality." *Southern Cross University Law Review* 2, 1–23.

Nussbaum, Martha C. 2006. "The Moral Status of Animals." *Chronicle of Higher Education*. Online.

Osvath, Mathias and Helena Osvath. 2008. "Chimpanzee (*Pan troglodytes*) and Orangutan (*Pan abelii*) Forethought: Self-Control and Pre-Experience in the Face of Future Tool Use." *Animal Cognition*. doi: 10.1007/s10071-008-0157-0.

Pagel, Mark. 2012. *Wired for Culture: Origins of the Human Social Mind*. NY: W.W. Norton.

Palmer, Alexandra. 2018. *Saving and Sacrificing: Ethical Questions in Orangutan Rehabilitation*. Ph.D. thesis, University College, London.

Panoz-Brown, Danielle, et al. 2018. "Replay of Episodic Memories in the Rat." *Current Biology* in press. doi: 10.1016/j.cub.2018.04.006.

Parker, Sue Taylor. 1999. "The Life History and Development of Great Apes in Comparative Perspective." 1999. *The Mentalities of Gorillas and Orangutans: Comparative Perspectives*. Sue Taylor Parker, Robert W. Mitchell, and H. Lyn Miles, eds. Cambridge: Cambridge University Press, 43–69.

Parker, Sue Taylor and Kathleen Rita Gibson, eds. 1990. *"Language" and Intelligence in Monkeys and Apes: Comparative Developmental Perspectives*. Cambridge: Cambridge University Press.

Parker, Sue Taylor and Michael McKinney. 1999b. *Origins of Intelligence: The Evolution of Cognitive Development in Monkeys, Apes, and Humans*. Baltimore: Johns Hopkins UP.

Parker, Sue Taylor, Robert Mitchell, and Maria L. Broccia. 1994. "Expanding Dimensions of the Self: Through the Looking Glass and Beyond." *Self-awareness in Animals and Humans: Developmental Perspectives*. Sue Taylor Parker, Robert W. Mitchell, and Maria L. Boccia, eds. Cambridge: Cambridge University Press, 3–19.

Parker, Sue Taylor, Robert W. Mitchell, and H. Lyn Miles, eds. 1999. *The Mentalities of Gorillas and Orangutans: Comparative Perspectives*. Cambridge: Cambridge University Press.

Parrish, Audrey E., et al. 2014. "Delay of Gratification by Orangutans (*Pongo pygmaeus*) in the Accumulation Task." *Journal of Comparative Psychology* 128.2, 209–214.

Patterson, Francine. 1978. "The Gestures of a Gorilla: Language Acquisition in Another Pongid." *Brain and Language* 5, 72–97.

Patterson, Francine and Wendy Gordon. 1993. "The Case for the Personhood of Gorillas." *The Great Ape Project: Equality Beyond Humanity*. Paola Cavalieri and Peter Singer, eds. NY: St. Martin's, 58–77.

Payne, Junaidi and Cede Prudente. 2008. *Orangutans: Behavior, Ecology and Conservation*. Cambridge, MA: MIT Press.

Penn, Derek C., et al. 2008. "Darwin's Mistake: Explaining the Discontinuity Between Human and Nonhuman Minds." *Behavioral Brain Sciences* 31, 109–78.

Peterson, Dale. 2011. *The Moral Lives of Animals*. NY: Bloomsbury Press.

Petter, Jean-Jacques and François Desbordes. 2010. *Primates of the World*. Translated by Robert Martin. Princeton: Princeton UP.

Pianka, Eric R. 1974. *Evolutionary Ecology*. NY: Harper and Row.

Pika, Simone, et al. 2019. "Wild Chimpanzees (*Pan troglodytes troglodytes*) Exploit Tortoises (*Kinixys erosa*) via Percussive Technology." *Scientific Reports* 9, 7661. https://doi.org/10.1038/s41598-019-43301-8.

Porter, Amy, et al. 2019. "Behavioral Responses Around Conspecific Corpses in Adult Eastern Gorillas (*Gorilla beringei spp.*)." *PeerJ* 7, e6655. http://doi.org/10.7717/peerj6655.

Povinelli, Daniel J. and Jennifer Vonk. 2003. "Chimpanzee Minds: Surprisingly Human?" *Trends in Cognitive Sciences* 7.4, 157–160.

Povinelli, Daniel J. and John G.H. Cant. 1995. "Arboreal Clambering and the Evolution of Self-conception." *The Quarterly Review of Biology* 70.4, 393–421.

Povinelli, Daniel J., et al. 1990. "Inferences About Guessing and Knowing by Chimpanzees (*Pan troglodytes*)." *Journal of Comparative Psychology* 104.3, 203–210.

Powell, Lauren E., et al. 2017. "Re-evaluating the Link Between Brain Size and Behavioural Ecology in Primates." *Proceedings of the Royal Society B* 284, 20171765.

Premack, David. 1971. "Language in Chimpanzee?" *Science* 172, 808–822.

Premack, Ann J. and David Premack. 1972. "Teaching Language to an Ape." *Scientific American* 227.4, 92–99.

Premack, David and Guy Woodruff. 1978. "Does the Chimpanzee Have a Theory of Mind?" *The Behavioral and Brain Sciences* 4, 515–526.

Premack, David and Ann J. Premack. 1994. "Levels of Causal Understanding in Chimpanzees and Children." *Cognition* 50, 347–362.

Preston, Stephanie D. and Frans B.M. de Waal. 2002. "Empathy: Its Ultimate and Proximate Bases." *Behavioral and Brain Sciences* 25, 1–71.

Preuschoft, Signe. 2000. "Primate Faces and Facial Expressions." *Social Research* 67.1, 245–271.

Prince-Hughes, Dawn. 1987. *Songs of the Gorilla Nation: My Journey through Autism*. Easton, PA: Harmony Press.

Prüefer, Kay, et al. 2012. "The Bonobo Genome Compared With Chimpanzee and Human Genomes." *Nature* 486, 527–531.

Prum, Richard O. 2017. *The Evolution of Beauty: How Darwin's Forgotten Theory of Mate Chioce Shapes the Animal World – And Us*. NY: Doubleday.

Prusinkiewicz, Przemyslaw, et al. 2009. "Control of Bud Activation by an Auxin Transport Switch." *PNAS* 106.41, 17431–17436. https://doi.org/10.1073/pnas.0906696106.

Purzycki, Benjamin Grant. 2005. "The Degree of Chimpanzee Theory of Mind and the Evolution of Modularity." *Transactions of the Nebraska Academy of Sciences* 30, 59–67.

Quinton, Anthony. 1967. "Karl Popper." *The Encyclopedia of Philosophy*. Volume 6. Paul Edwards, ed. NY: Macmillan, 398–401.

Rachels, James. 1990. *Created from Animals: The Moral Implications of Darwinism*. Oxford: OUP.

Rachels, James. 1993. "Why Darwinians Should Support Equal Treatment for Other Great Apes." *The Great Ape Project: Equality Beyond Humanity*. Paola Cavalieri and Peter Singer, eds. NY: St. Martin's, 152–157.

Rachels, James, ed. 1998. *Ethical Theory 1: The Question of Objectivity*. Oxford: OUP.
Radhakrishna, Sindhu and Dale Jamieson. 2018. "Liberating Primatology." *Journal of Bioscience* 43.1, 3–8.
Regan, Tom. 1983; 2004. *The Case for Animal Rights*. Berkeley: U California P.
Regan, Tom. 1993. "Ill-gotten Gains." *The Great Ape Project: Equality Beyond Humanity*. Paola Cavalieri and Peter Singer, eds. NY: St. Martin's, 194–205.
Reiserer, Randall S., et al. 2018. "Seed Ingestion and Germination in Rattlesnakes: Overlooked Agents of Rescue and Secondary Dispersal." *Proceedings of the Royal Society B* 285, 20172755. doi: 10.1098/rspb.2017.2755.
Richter, Daniel, et al. 2017. "The Age of the *Homo Sapiens* Fossils from Jebel Irhoud, Morocco, and the Origins of the Middle Stone Age." *Nature* 546, 293–296.
Rijksen, Herman and Erik Meijaard. 1999. *Our Vanishing Relative: The Status of Wild Oran-gutans at the Close of the Twentieth Century*. Springer/Tropenbos.
Riley, Erin P. 2006. "Ethnoprimatology: Toward Reconciliation of Biological and Cultural Anthropology." *Ecological and Environmental Anthropology* 2.2, 75–86.
Roffman, Itai, et al. 2015. "Preparation and Use of Varied Natural Tools for Extractive Foraging by Bonobos (*Pan Paniscus*)." *American Journal of Physical Anthropology* 158.1, 78–91. https://doi.org/10.1002/ajpa.22778.
Rollin, Bernard E. 1993. "The Ascent of Apes – Broadening the Moral Community." *The Great Ape Project: Equality Beyond Humanity*. Paola Cavalieri and Peter Singer, eds. NY: St. Martin's, 206–219.
Rollin, Bernard E. 2006. *Animal Rights and Human Morality*, third edition. Amherst, NY: Prometheus Books.
Rollin, Bernard E. 2009. "The Moral Status of Animals and Their Use as Experimental Subjects." *A Companion to Bioethics*, second edition. Helga Kuhse and Peter Singer, eds. Malden, MA: Blackwell, 495–509.
Román, Valeria. 2015. "Argentina Grants an Orangutan human-like Rights." *Scientific American*. https://www.scientificamerican.com/article/argentina-grants-an-orangutan-human-like-rights/.
Rosati, Alexandra G. 2017. "The Evolution of Primate Executive Function: From Response Control to Strategic Decision-Making." *Evolution of Nervous Systems*, second ed. J. Kaas and L. Krubitzer, eds. Cambridge, MA: Harvard UP, 423–437.
Rosati, Alexandra G., et al. 2016. "Rhesus Monkeys Show Human-like Changes in Gaze Following Across the Lifespan." *Proceedings of the Royal Society B* 283, 20160376.
Rose, Anthony L. 1994. "Description and Analysis of Profound Interspecies Events." *Proceedings of the XVth Congress of International Primatological Society*, Bali, Indonesia.
Rose, Anthony L. 1995. "Orangutans, Science, and Collective Reality." *The Neglected Ape*. Ronald D. Nadler, et al., eds. NY: Plenum, 29–40.
Rose, Anthony L. 2011. "Bonding, Biophilia, Biosynergy, and the Future of Primates in the Wild." *American Journal of Primatology* 73, 245–252.
Ross, Marina Davila, et al. 2007. "Rapid Facial Mimicry in Orangutan Play." *Biology Letters* 4, 27–30. doi: 10.1098/rsbl.2007.0535.

Roth, Gerhard and Ursula Dicke. 2005. "Evolution of the Brain and Intelligence." *Trends in Cognitive Science* 9.5, 250–257. doi: 10.1016/j.tics.2005.03.005.

Rowlands, Mark. 2012. *Can Animals Be Moral?* Oxford: OUP.

Rowlands, Mark. 2019. *Can Animals Be Persons?* Oxford: OUP.

Russon, Anne E. 1999. *The Mentalities of Gorillas and Orangutans: Comparative Perspectives*. Sue Taylor Parker, Robert W. Mitchell, and H. Lyn Miles, eds. Cambridge: Cambridge University Press, 117–146.

Russon, Anne E. 2004a. *Orangutans: Wizards of the Rainforest*. NY: Firefly Books.

Russon, Anne E. 2004b. "Evolutionary Reconstruction of Great Ape Intelligence." *The Evolution of Thought: Evolutionary Origins of Great Ape Intelligence*. Anne E. Russon and David R. Begun, eds. Cambridge: Cambridge University Press, 1–14.

Russon, Anne E. and Biruté Galdikas. 1995. "Initiation and Tool Use in Rehabilitant Orangutans." *The Neglected Ape*. Ronald D. Nadler, et al., eds. NY: Plenum, 191–197.

Russon, Anne E. and David R. Begun. 2004a. "Evolutionary Origins of Great Ape Intelligence: An Integrated View." *The Evolution of Great Ape Intelligence*. A. E. Russon and D. R. Begun, eds. Cambridge: Cambridge University Press, 353–368.

Russon, Anne E. and David Begun, eds. 2004b. *The Evolution of Thought: Evolutionary Origins of Great Ape Intelligence*. Cambridge: Cambridge University Press.

Russon, Anne E. and Kim A. Bard. 1996. "Exploring the Minds of the Great Apes: Issues and Controversies." *Reaching Into Thought: The Minds of the Great Apes*. Anne E. Russon, Kim A. Bard, and Sue Taylor Parker, eds. Cambridge: Cambridge University Press, 1–20.

Russon, Anne E., Kim A. Bard, and Sue Taylor Parker, eds. 1996. *Reaching Into Thought: The Minds of the Great Apes*. Cambridge: Cambridge University Press.

Russon, Anne E., et al. 2008. "Geographic Variation in Orangutan Diets." *Orangutans: Geographic Variation in Behavioral Ecology and Conservation*. Serge A. Wich, ed. Oxford: OUP, 135–156.

Sagoff, Mark. 2013. "Animal Liberation and Environmental Ethics: Bad Marriage, Quick Divorce." *Reflecting on Nature*, second edition. Lori Gruen, Dale Jamieson, and Christopher Schlottmann, eds. NY: Oxford University Press, 160–166.

Sale, Kirkpatrick. 2006. *After Eden: The Evolution of Human Domination*. Durham, NC: Duke UP.

Saripan, Hartini, et al. 2016. "Are Robots Human? A Review of the Legal Personality Model." *World Applied Sciences Journal* 34.6, 824–831.

Savage-Rumbaugh, E. Sue and Duane Rumbaugh. 1978. "Symbolization, Language, and Chimpanzees: A Theoretical Re-evaluation Based on Initial Language Acquisition Processes in Four Young Pan troglodytes." *Brain and Language* 6, 265–300.

Savage-Rumbaugh, Sue, et al. 2000. "Ape Consciousness-Human Consciousness: A Perspective Informed by Language and Culture." *American Zoologist* 40, 910–921.

Savage-Rumbaugh, Sue, et al. 2007. "Welfare of Apes in Captive Environments: Comments on, and by, a Specific Group of Apes." *Journal of Applied Animal Welfare Science* 10.1, 7–19.

Savage-Rumbaugh, Sue, et al. 2018. "The Fully Conscious Ape." *International Journal of Comparative Psychology* 31. Online.

Scassellati, Brian. 2002. "Theory of Mind for a Humanoid Robot." *Autonomous Robots* 12.1, 13–24.

Schmitt, Richard. 1967. "Phenomenology." *The Encyclopedia of Philosophy*. Volume 6. Paul Edwards, ed. NY: Macmillan, 135–151.

Schopenhauer, Arthur. 1840. *On the Basis of Morality*. Trans. E.F.J. Payne. Indianapolis: Bobbs Merrill, 1965.

Schuppli, Caroline and Carel P. van Schaik. 2019. "Animal Cultures: How We've Only Seen the Tip of the Iceberg." *Evolutionary Human Sciences* 1, e2, 1–13. doi: 10.1017/ehs.2019.1.

Schwartz, Jeffrey H. 1987. *The Red Ape: Orang-utans and Human Origins*. Boston: Houghton Mifflin.

Scott-Phillips, Thomas C. 2016. "Meaning in Great Ape Communication." *Animal Cognition* 19, 233–238.

Semendeferi, Katerina. 1999. "The Frontal Lobes of the Great Apes With a Focus on the Gorilla and the Orangutan." *The Mentalities of Gorillas and Orangutans: Comparative Perspectives*. Sue Taylor Parker, Robert W. Mitchell, and H. Lyn Miles, eds. Cambridge: Cambridge University Press, 70–95.

Seyfarth, Robert M. and Dorothy L. Cheney. 2013. "The Primate Mind before Tools, Language, and Culture." *Evolution of Mind, Brain, and Culture*. Gary Hatfield and Holly Pittman, eds. Philadelphia: U Pennsylvania P, 105–121.

Shapiro, Gary. 1982. "Sign Acquisition in a Home-reared/Free-ranging Orangutan: Comparisons with Other Signing Apes." *American Journal of Primatology* 3, 121–129.

Shapiro, Gary. 1985. *Factors Influencing Sign Learning Performance in Four Juvenile Orangutans*. Ph.D. Dissertation, University of Oklahoma, Dept. of Zoology. University Microfilms International, Ann Arbor, MI.

Shapiro, Gary and Biruté Galdikas. 1999. "Early Sign Performance in a Free-ranging, Adult Orangutan." *The Mentalities of Gorillas and Orangutans: Comparative Perspectives*. Sue Taylor Parker, Robert W. Mitchell and H. Lyn Miles, eds. Cambridge: Cambridge University Press, 265–279.

Shapiro, Paul. 2006. "Moral Agency in Other Animals." *Theoretical Medicine and Bioethics* 27, 357–373. doi: 10.1007/s11017-006-9010-0.

Sharma, Rajiv and Judges Lok Pal Singh. 2018. High Court of Uttarakhand at Nainital. Writ petition No. 43 of 2014.

Shi, Lei, et al. 2019. "Transgenic Rhesus Monkeys Carrying the Human *MCPH1* Gene Copies Show Human-like Neotony of Brain Development." *National Science Review* 6, 480–493. doi: 10.1093/nsr/nwz043.

Simpson, G. Gaylord. 1965. *The Geography of Evolution*. NY: Capricorn Books.

Singer, Peter. 1975. *Animal Liberation*. NY: Ecco/Harper Collins, 2009.

Singer, Peter. 2015. "There is No Good Reason to Keep Apes in Prison." *Wired*, 26 May 2015. Online.

Smith, Adam. 1897. *The Theory of Moral Sentiments*. 1759. *British Moralists*. Volume 1. L.A. Selby-Bigge, ed. Oxford: Clarendon P, 251–336.

Smith, David Livingstone and Ioana Panaitiu. 2015. "Aping the Human Essence: Simianization as Dehumanization." *Simianization: Apes, Gender, Class, and Race*. Wulf D. Hund, Charles W. Mills, and Silvia Sebastiani, eds. Berlin: LIT Verlag, 77–103.

Smith, Joshua J. 2014. "Human-Animal Relationships in Zoo-Housed Orangutans (*P. abelii*) and Gorillas (*G.g. gorilla*): The Effects of Familiarity." *American Journal of Primatology* 76, 942–955.

Smith, Julie A. and Robert W. Mitchell, eds. 2012a. *Experiencing Animal Minds*. NY: Columbia University Press.

Smith, Julie A. and Robert W. Mitchell. 2012b. "Animal Ethics and Animals' Minds." *Experiencing Animal Minds*. Julie A. Smith and Robert W. Mitchell, eds. NY: Columbia University Press, 333–355.

Snow, C.P. 1959. "The Two Cultures." *The Two Cultures*. Stefan Collini, ed. Cambridge: Cambridge University Press, 1998.

Sousa, André M.M., et al. 2017. "Molecular and Cellular Reorganization of Neural Circuits in the Human Lineage." *Science* 358.6366, 1027–1032. doi: 10.1126/science.aan3456.

Sowards, Stacey K. 2006. "Identification Through Orangutans: Destabilizing the Nature/Culture Dualism." *Ethics and the Environment* 11.2, 45–61.

Spehar, Stephanie N., et al. 2018. "Orangutans Venture Out of the Rainforest and Into the Anthropocene." *Science Advances* 4, e1701422. Online.

Spence, Caroline E., et al. 2017. "Theory of Animal Mind: Human Nature or Experimental Artefact." *Trends in Cognitive Sciences* 21.5, 333–343. doi: 10.1016/j.tics.2017.02.003.

Staes, Nicky, et al. 2017. "*FOXP2* Variation in Great Ape Populations Offers Insight Into the Evolution of Communication Skills." *Scientific Reports* 7, 16866. doi: 10.1038/s41598-017-16844-x.

Stewart, Fiona Anne. 2011. *The Evolution of Shelter: Ecology and Ethology of Chimpanzee Nest Building*. Ph.D. dissertation. University of Cambridge, UK, Corpus Christi College. Online.

Stotz, Karola. 2017. "Why Developmental Niche Construction is Not Selective Niche Construction: And Why It Matters." *Interface Focus* 7, 20160157. Online.

Strum, Shirley C. 1987. *Almost Human: A Journey Into the World of Baboons*. NY: Norton.

Suarez, Susan and Gordon Gallup, Jr. 1981. "Self-Recognition in Chimpanzee and Orangutans, But Not Gorillas." *Journal of Human Evolution* 10, 175–188.

Swartz, Karyl. 1997. "What is Mirror-Recognition, and What is it Not?" *Annals of the New York Academy of Sciences* 818, 65–71.

Swartz, Karyl B., et al. 1999. "Comparative Aspects of Mirror Self-recognition in Great Apes." *The Mentalities of Gorillas and Orangutans*. Sue Taylor Parker, Robert Mitchell, and H. Lyn Miles, eds. Cambridge: Cambridge University Press, 283–294.

Tague, Gregory F. 2014. *Making Mind: Moral Sense and Consciousness*. Amsterdam: Rodopi.

Tague, Gregory F. 2016. *Evolution and Human Culture*. Leiden: Brill.

Tague, Gregory F. 2018. *Art and Adaptability: Consciousness and Cognitive Culture*. Leiden: Brill.

Tague, Gregory F. 2019. "Story of Jimmy." YouTube. https://www.youtube.com/watch?v=8MVxWKtqLZw&t=3s.

Tan, Jingzhi, et al. 2017. "Bonobos Respond Prosocially Toward Members of Other Groups." *Scientific Reports* 7, 14733. doi: 10.1038/s41598-017-15320-w.

Tarszisz, Esther, et al. 2018. “An Ecophysiologically Informed Model of Seed Dispersal by Orangutans: Linking Animal Movement With Gut Passage Across Time and Space.” *Conservation Physiology* 6, 1. doi: 10.1093/conphys/coy013.

Thompson, Shawn. 2002. *Letters from Prison: Felons Write about the Struggle for Life and Sanity Behind Bars*. Harper Collins.

Thompson, Shawn. 2010. *The Intimate Ape: Orangutans and the Secret Life of a Vanishing Species*. Kensington.

Thompson, Shawn. 2019a. “Supporting Ape Rights: Finding the Right Fit Between Science and the Law.” *ASEBL Journal* 14.1, 3–25. Online.

Thompson, Shawn. 2019b. “How to Resist Anxiety When You Write an Overly Ambitious Article about Science, the Law and Animal Rights.” *ASEBL Journal* 14.1. Online.

Tinbergen, Niko. 1963. “On the Aims and Methods of Ethology.” *Zeitschrift für Tierpsychologie* 20, 410–433. Online.

Tomasello, Michael. 2014. *A Natural History of Human Thinking*. Cambridge, MA: Harvard UP.

Trivers, Robert. 1971. “The Evolution of Reciprocal Altruism.” *The Quarterly Review of Biology* 46.1, 35–57.

Trivers, Robert. 2002. *Natural Selection and Social Theory*. Oxford: OUP.

Trivers, Robert. 2011. *The Folly of Fools*. NY: Basic Books.

UNEP. 2005. The Kinshasa Declaration on Great Apes. UNESCO. https://www.cites.org/sites/default/files/eng/news/sundry/2005/kinshasa_declaration.pdf.

United Nations. 1982. World Charter for Nature. A/RES/37/7.

van Casteren, Adam, et al. 2012. “Nest-building Orangutans Demonstrate Engineering Know-how to Produce Safe, Comfortable Beds.” *PNAS* 109.18, 6873–6877. doi: 10.1073/pnas/1200902109.

van der Weele, Cor. 2011. “Empathy’s Purity, Sympathy’s Complexities; De Waal, Darwin and Adam Smith.” *Biology and Philosophy* 26, 583–593.

van Hooff, Jan A. R. A. M. and Bas Lukkenaar. 2015. “Captive Chimpanzee Takes Down a Drone: Tool Use Toward a Flying Object.” *Primates*. Online.

van Leeuwen, Edwin J.C., et al. 2014. “A Group-specific Arbitrary Tradition in Chimpanzees (*Pan troglodytes*).” *Animal Cognition* 17, 1421–1425.

van Leeuwen, Edwin J.C., et al. 2016. “Chimpanzees’ Responses to the Dead Body of a 9-Year-Old Group Member.” *American Journal of Primatology*. doi: 10.1002/ajp.22560. Online.

van Leeuwen, Edwin J.C., et al. 2017. “Tool Use for Corpse Cleaning in Chimpanzees.” *Scientific Reports* 7, 44091. doi: 10.1038/srep44091.

van Schaik, Carel P. and Jan A.R.A.M. van Hooff. 1996. “Toward an Understanding of the Orangutan’s Social System.” *Great Ape Societies*. William C. McGrew, Linda Marchant, and Toshisada Nishida, eds. Cambridge: Cambridge University Press, 3–15.

van Schaik, Carel P., et al. 1996. “Manufacture and Use of Tools in Wild Sumatran Orangutans: Implications for Human Evolution.” *Naturwissenschaften* 83, 186–188.

van Schaik, Carel P., et al. 2003. “Orangutan Cultures and the Evolution of Material Culture.” *Science* 299, 102–105.

van Schaik, Carel P., et al. 2006. "Innovation in Wild Bornean Orangutans (*Pongo pygmaeus wurmbii*)." *Behaviour* 143, 839–876.

van Schaik, Carel P., et al. 2013. "Wild Orangutan Males Plan and Communicate Their Travel Direction One Day in Advance." *Plos One* 8.9, e74896. doi: 10.1371/journal.pone.0074896.

van Schaik, Carel P., et al. 2016. "The Reluctant Innovator: Orangutans and the Phylogeny of Creativity." *Philosophical Transactions of the Royal Society B* 371, 20150183. doi.org/10.1098/rstb.2015.0183.

van Wyhe, John and Peter C. Kjaergaard. 2015. "Going the Whole Orang: Darwin, Wallace and the Natural History of Orangutans." *Studies in History and Philosophy of Biological and Biomedical Sciences* 51, 53–63. doi: 10.1016/j.shpsc.2015.02.006.

Varela, Francisco J. 1992. "Whence Perceptual Meaning? A Cartography of Current Ideas." *Understanding Origins*. Springer, 235–263.

Varner, Gary E. 2012. *Personhood, Ethics, and Animal Cognition*. Oxford: OUP.

Viola, Marco. 2017. "Commentary: Constructing Nonhuman Animal Emotion." *Frontiers in Psychology* 8, 2070. doi: 10.3389/fpsyg.2017.02070.

Voigt, Maria, et al. 2018. "Global Demand for Natural Resources Eliminated More Than 100,000 Bornean Orangutan." *Current Biology* 28.5, 761–769e5. doi: 10.1016/j.cub.2018.01.053.

Vonk, Jennifer. 2014. "Quantity Matching by an Orangutan (*Pongo abelii*)." *Animal Cognition* 17, 297–306. doi: 10.1007/s10071-013-0662-7.

Vonk, Jennifer and Suzanne E. McDonald. 2004. "Levels of Abstraction in Orangutan (*Pongo abelii*) Categorization." *Journal of Comparative Psychology* 118.1, 3–13. doi: 10.1037/0735-7036.118.1.3.

von Rohr, Claudia Rudolf, et al. 2015. "Chimpanzees' Bystander Reactions to Infanticide: An Evolutionary Precursor of Social Norms?" *Human Nature* 26, 143–160.

von Uexküll, Jakob. 2010. *A Foray Into the World of Animals and Humans*. Joseph D. O'Neil, trans. Introduction by Dorion Sagan. Minneapolis: U Minnesota P.

Waller, Sara. 2012. "Science of the Monkey Mind: Primate Penchants and Human Pursuits." *Experiencing Animal Minds*. Julie A. Smith and Robert W. Mitchell, eds. NY: Columbia University Press, 79–94.

Warneken, Felix. 2013. "The Origins of Human Cooperation from a Developmental and Comparative Perspective." *Evolution of Mind, Brain, and Culture*. Gary Hatfield and Holly Pittman, eds. Philadelphia: University of Pennsylvania Museum of Archaeology and Anthropology, 149–168.

Webb, Christine, Peter Woodford, and Elise Huchard. 2019. "Animal Ethics and Behavioral Science: An Overdue Discussion?" *BioScience*. Oxford University Press, in press.

Wechkin, S., et al. 1964. "Shock to a Conspecific as an Aversive Stimulus." *Psychonomic Science* 1, 17–18.

Weisberg, Zipporah. 2015. "Animal Agency: What It Is, What It Isn't, and How It Can Be Realized." *Animal Ethics and Philosophy: Questioning the Orthodoxy*. Elisa Aaltola and John Hadley, eds. London: Rowman and Littlefield, 63–80.

Weisberg, Zipporah. 2019. "The Problem with the Personhood Argument." *ASEBL Journal* 14.1, 33–36. Online.

Weiss, Alexander and James E. King. 2015. "Great Ape Origins of Personality Maturation and Sex Differences: A Study of Orangutans and Chimpanzees." *Journal of Personality and Social Psychology* 108.4, 648–664. doi: 10.1037/pspp0000022.

Wenz, Peter S. 1993. "Ecology and Morality." *Moral Controversies*. Stephen Jay Gould, ed. Belmont, CA: Wadsworth, 545–550.

Whiten, Andrew. 1996. "Imitation, Pretense, and Mindreading: Secondary Representation in Comparative Primatology and Developmental Psychology?" *Reaching Into Thought: The Minds of the Great Apes*. Anne E. Russon, Kim A. Bard, and Sue Taylor Parker, eds. Cambridge: Cambridge University Press, 300–324.

Whiten, Andrew. 2010. "Ape Behavior and the Origins of Human Culture." *Mind the Gap: Tracing the Origins of Human Universals*. Peter M. Kappeler and Joan B. Silk, eds. Heidelberg: Springer, 429–450.

Whiten, Andrew, et al. 1999. "Culture in Chimpanzees." *Nature* 399, 682–685.

Whiten, Andrew, et al. 2009. "Emulation, Imitation, Over-imitation and the Scope of Culture for Child and Chimpanzee." *Philosophical Transactions of the Royal Society B* 364, 2417–2428.

Wich, Serge, et al. 2008. "Distribution and Conservation Status of the Orang-utan (*Pongo* spp.) on Borneo and Sumatra: How Many Remain?" Fauna & Flora International. *Oryx* 42.3, 329–339.

Wildman, Derek E., et al. 2003. "Implications of Natural Selection in Shaping 99.4% Nonsynonymous DNA Identity Between Humans and Chimpanzees: Enlarging Genus *Homo*." *PNAS* 100.12, 7181–7188.

Williams, George C. 1966. *Adaptation and Natural Selection*. Princeton: Princeton UP.

Williams, Wayne. 2015. "Enchanted Worlds and Animal Others." *Animal Ethics and Philosophy: Questioning the Orthodoxy*. Elisa Aaltola and John Hadley, eds. London: Rowman and Littlefield, 83–98.

Wilson, Edward O. 1975. *Sociobiology: The New Synthesis*. Cambridge, MA: Harvard UP.

Wilson, Edward O. 1984. *Biophilia: The Human Bond With Other Species*. Cambridge, MA: Harvard UP.

Wilson, Edward O. 2017. *The Origins of Creativity*. NY: Liveright.

Wise, Steven M. 2000. *Rattling the Cage: Toward Legal Rights for Animals*. Cambridge, MA: Perseus.

Wise, Steven M. 2002. *Drawing the Line: Science and the Case for Animal Rights*. Cambridge, MA: Perseus.

Wise, Steven M. 2017. "Introduction to Animal Law Book." *Syracuse Law Review* 67.7, 7–30.

Wolf, Julie, et al. 2017. "Revised Methane Emission Factors and Spatially Distributed Annual Carbon Fluxes for Global Livestock." *Carbon Balance and Management* 12, 16. doi: 10.1186/s13021-017-0084-y.

Wolf, Wouter and Michael Tomasello. 2019. "Visually Attending to a Video Together Facilitates Great Ape Social Closeness." *Proceedings of the Royal Society B* 286, 20190488. doi: 10.1098/rspb.2019.0488.

Wright, Justin P. and Clive G. Jones. 2006. "The Concept of Organisms as Ecosystem Engineers Ten Years On: Progress, Limitations, and Challenges." *BioScience* 56.3, 203–209.

WWF. 2018. *Living Planet Report – 2018: Aiming Higher*. M. Grooten and R. E. A. Almond, eds. Gland, Switzerland: WWF.

Yamagiwa, Juichi. 2004. "Diet and Foraging of the Great Apes: Ecological Constraints on Their Social Organizations and Implications for Their Divergence." *The Evolution of Thought: Evolutionary Origins of Great Ape Intelligence*. Anne E. Russon and David R. Begun, eds. Cambridge: Cambridge University Press, 210–233.

Yamakoshi, Gen. 2004. "Evolution of Complex Feeding Techniques in Primates: Is This the Origin of Great Ape Intelligence?" *The Evolution of Thought: Evolutionary Origins of Great Ape Intelligence*. Anne E. Russon and David R. Begun, eds. Cambridge: Cambridge University Press, 140–171.

Yerkes, Robert M. 1916. *The Mental Life of Monkeys and Apes: A Study of Ideational Behavior*. Boston: Henry Holt.

Index

About the Author

Gregory F. Tague, PhD (1998), New York University, is professor of English/Interdisciplinary Studies and founder and senior developer of The Evolutionary Studies Collaborative at St. Francis College, NY. He is also the founder and organizer of a number of Darwin-inspired Moral Sense Colloquia and other multidisciplinary events. Recent, relevant books include *Art and Adaptability: Consciousness and Cognitive Culture* (Brill, 2018), *Evolution and Human Culture* (Brill, 2016), and *Making Mind: Moral Sense and Consciousness* (Rodopi, 2014). Tague has written and edited a number of other academic books, such as *Character and Consciousness* (2005) and *Origins of English Dramatic Modernism* (2010). He is the founding editor of the peer-reviewed *ASEBL Journal* (ethics/arts/evolution) and general editor of the Bibliotekos literary website. Currently, he is the general editor of *Literary Veganism: An Online Journal*. Tague has also edited five literary anthologies, notably, *Being Human* (2012) and *Puzzles of Faith and Patterns of Doubt* (2013).

www.ingramcontent.com/pod-product-compliance
Ingram Content Group UK Ltd.
Pitfield, Milton Keynes, MK11 3LW, UK
UKHW012119150926
475560UK00002B/176

9 781793 619709